TEACHER RESOURCES

Human Systems Interactions

Full Option Science System
Developed at the Lawrence Hall of Science, University of California, Berkeley
Published and Distributed by Delta Education

FOSS Lawrence Hall of Science Team
Larry Malone and Linda De Lucchi, FOSS Project Codirectors and Lead Developers Jessica Penchos, Middle School Coordinator; Kathy Long, FOSS Assessment Director; David Lippman, Program Manager; Carol Sevilla, Publications Design Coordinator; Susan Stanley, Graphic Production; Rose Craig, Illustrator
FOSS Curriculum Developers: Alan Gould, Teri Lawson, Ann Moriarty, Virginia Reid, Terry Shaw, Joanna Snyder
FOSS Technology Developers: Susan Ketchner, Arzu Orgad
FOSS Multimedia Team: Kate Jordan, Senior Multimedia; Christopher Keller, Multimedia Producer; Jonathan Segal, Designer; Christopher Cianciarulo, Designer; Dan Bluestein, Programmer; Shan Jiang, Programmer

Delta Education Team
Bonnie A. Piotrowski, Editorial Director, Elementary Science
Project Team: Jennifer Apt, Sandra Burke, Mary Connell, Joann Hoy, Angela Miccinello, Jennifer Staltare

Content Reviewer
Rocco Carsia, Associate Professor
Department of Cell Biology
School of Osteopathic Medicine
Rowan University, New Jersey

Thank you to all FOSS Middle School Revision Trial Teachers and District Coordinators
Frances Amojioyi, Lincoln Middle School, Alameda, CA; Dean Anderson, Organized trials for Boston Public Schools, Boston, MA; Thomas Archer, Organized trials for ESD 112, Vancouver, WA; Lauresa Baker, Lincoln Middle School, Alameda, CA; Bobbi Anne Barnowsky, Canyon Middle School, Castro Valley, CA; Christine Bertko, St. Finn Barr Catholic School, San Francisco, CA; Stephanie Billinge, James P. Timilty Middle School, Roxbury, MA; Jerry Breton, Ingleside Middle School, Phoenix, AZ; Robert Cho, Timilty Middle School, Boston, MA; Susan Cohen, Cherokee Heights Middle School, Madison, WI; Malcolm Davis, Canyon Middle School, Castro Valley, CA; Marilyn Decker, Organized trials for Milton PS, Milton, MA; Jenny Ernst, Park Day School, Oakland, CA; Marianne Floyd, Spanaway Middle School, Spanaway, WA; Sarah Kathryn Gessford, Journeys School, Jackson, WY; Charles Hardin, Prairie Point Middle School, Cedar Rapids, IA; Jennifer Hartigan, Lincoln Middle School, Alameda, CA; Sheila Holland, TechBoston Academy, Boston, MA; Nicole Hoyceanyls, Charles S. Pierce Middle School, Milton, MA; Bruce Kamerer, Donald McKay K-8 School, East Boston, MA; Carmen Saele Kardokus, Reeves Middle School, Olympia, WA; Janey Kaufman, Organized trials for Scottsdale USD, Scottsdale, AZ; Erica Larson, Organized trials for Grant Wood AEA, Cedar Rapids, IA; Lindsay Lodholz, O'Keeffe Middle School, Madison, WI; Robert Mattisinko, Chaparral High School, Scottsdale, AZ; Brenda McGurk, Prairie Point Middle School, Cedar Rapids, IA; Tim Miller, Mountainside Middle School, Scottsdale, AZ; Thomas Miro, Lincoln Middle School, Alameda, CA; Spencer Nedved, Frontier Middle School, Vancouver, WA; Joslyn Olsen, Lincoln Middle School, Alameda, CA; Stephanie Ovechka, Cedarcrest Middle School, Spanaway, WA; Barbara Reinert, Copper Ridge School, Scottsdale, AZ; Stephen Ramos, Lincoln Middle School, Alameda, CA; Gina Rutenbeck, Prairie Point Middle School, Cedar Rapids, IA; John Sheridan, Boston Public Schools (Boston Schoolyard Initiative), Boston, MA; Barbara Simon, Timilty Middle School, Boston, MA; Lise Simpson, Alcott Middle School, Norman, OK; Autumn Stevick, Thurgood Marshall Middle School, Olympia, WA; Ted Stoeckley, Hall Middle School, Larkspur, CA; Lesli Taschwer, Organized trials for Madison SD, Madison, WI; Paula Warner, Alcott Middle School, Norman, OK; Darren T. Wells, James P. Timilty Middle School, Boston, MA; Kristin White, Frontier Middle School, Vancouver, WA

Photo Credits: © MichaelTaylor3d/Shutterstock (cover); © Delta Education

Published and Distributed by Delta Education, a member of the School Specialty Family
The FOSS program was developed in part with the support of the National Science Foundation grant nos. ESI-9553600 and ESI-0242510. However, any opinions, findings, conclusions, statements, and recommendations expressed herein are those of the authors and do not necessarily reflect the views of NSF. FOSSmap was developed in collaboration between the BEAR Center at UC Berkeley and FOSS at the Lawrence Hall of Science. Score analysis is done through the BEAR Center Scoring Engine.

Copyright © 2017 by The Regents of the University of California

All rights reserved. Any part of this work (other than duplication masters) may not be reproduced or transmitted in any form or by any means, electronic or mechanical, including photocopying and recording, or by an information storage or retrieval system without permission of the University of California. For permission please write to: FOSS Project, Lawrence Hall of Science, University of California, Berkeley, CA 94720.

Human Systems Interactions — Teacher Toolkit, 1465646
Teacher Resources, 1465697
978-1-62571-209-7
Printing 1 – 3/2016
Patterson Printing, Benton Harbor, MI

TEACHER RESOURCES

TABLE OF CONTENTS

FOSS Program Goals A1
Science Notebooks in Middle School B1
Science-Centered Language Development in Middle School C1
FOSSweb and Technology D1
Science Notebook Masters 1–13
Teacher Masters . . . . A–R
Assessment Masters
Assessment Charts 1–9
Entry-Level Survey 1–2
Investigations 1–2 I-Check 1–4
Investigation 3 I-Check .1–4
Posttest 1–4

Notebook Answers

This document, *Teacher Resources*, is one of three parts of the *FOSS Teacher Toolkit* for this course. The chapters in *Teacher Resources* are all available as PDFs on FOSSweb.

The other parts of the course *Teacher Toolkit* are the *Investigations Guide* and a copy of the *FOSS Science Resources* student book containing original readings for this course.

The spiral-bound *Investigations Guide* contains these chapters.

- Overview
- Framework and NGSS
- Materials
- Investigations
- Assessment

The *Teacher Toolkit* is the most important part of the FOSS Program. It is here that all the wisdom and experience contributed by hundreds of educators has been assembled. Everything we know about the content of the course, how to teach the subject, and the resources that will assist the effort are presented here.

FOSS Program Goals

FOSS Program Goals

Contents

Introduction A1

Goals of the FOSS Program ... A2

Bridging Research
into Practice A5

FOSS Next Generation
K–8 Scope and Sequence A8

INTRODUCTION

The Full Option Science System™ has evolved from a philosophy of teaching and learning at the Lawrence Hall of Science that has guided the development of successful active-learning science curricula for more than 40 years. The FOSS Program bridges research and practice by providing tools and strategies to engage students and teachers in enduring experiences that lead to deeper understanding of the natural and designed worlds.

Science is a creative and analytic enterprise, made active by our human capacity to think. Scientific knowledge advances when scientists observe objects and events, think about how they relate to what is known, test their ideas in logical ways, and generate explanations that integrate the new information into understanding of the natural world. Engineers apply that understanding to solve real-world problems. Thus the scientific enterprise is both what we know (content knowledge) and how we come to know it (practices). Science is a discovery activity, a process for producing new knowledge.

The best way for students to appreciate the scientific enterprise, learn important scientific and engineering concepts, and develop the ability to think well is to actively participate in scientific practices through their own investigations and analyses. FOSS was created to engage students and teachers with meaningful experiences in the natural and designed worlds.

Full Option Science System

FOSS Program Goals

GOALS OF THE FOSS PROGRAM

FOSS has set out to achieve three important goals: scientific literacy, instructional efficiency, and systemic reform.

Scientific Literacy

FOSS provides all students with science experiences that are appropriate to students' cognitive development and prior experiences. It provides a foundation for more advanced understanding of core science ideas that are organized in thoughtfully designed learning progressions and prepares students for life in an increasingly complex scientific and technological world.

The National Research Council (NRC) in *A Framework for K–12 Science Education: Practices, Crosscutting Concepts, and Core Ideas* and the American Association for the Advancement of Science (AAAS) in *Benchmarks for Scientific Literacy* have described the characteristics of scientific literacy:

- Familiarity with the natural world, its diversity, and its interdependence.

- Understanding the disciplinary core ideas and the crosscutting concepts of science, such as patterns; cause and effect; scale, proportion, and quantity; systems and system models; energy and matter—flows, cycles, and conservation; structure and function; and stability and change.

- Knowing that science and engineering, technology, and mathematics are interdependent human enterprises and, as such, have implied strengths and limitations.

- Ability to reason scientifically.

- Using scientific knowledge and scientific and engineering practices for personal and social purposes.

The FOSS Program design is based on learning progressions that provide students with opportunities to investigate core ideas in science in increasingly complex ways over time. FOSS starts with the intuitive ideas that primary students bring with them and provides experiences that allow students to develop more sophisticated understanding as they grow through the grades. Cognitive research tells us that learning involves individuals in actively constructing schemata to organize new information and to relate and incorporate the new understanding into established knowledge. What sets experts apart from novices is that

Full Option Science System

experts in a discipline have extensive knowledge that is effectively organized into structured schemata to promote thinking. Novices have disconnected ideas about a topic that are difficult to retrieve and use. Through internal processes to establish schemata and through social processes of interacting with peers and adults, students construct understanding of the natural world and their relationship to it.

The target goal for FOSS students is to know and use scientific explanations of the natural world and the designed world; to understand the nature and development of scientific knowledge and technological capabilities; and to participate productively in scientific and engineering practices.

Instructional Efficiency

FOSS provides all teachers with a complete, cohesive, flexible, easy-to-use science program that reflects current research on teaching and learning, including student discourse, argumentation, writing to learn, and reflective thinking, as well as teacher use of formative assessment to guide instruction. The FOSS Program uses effective instructional methodologies, including active learning, scientific practices, focus questions to guide inquiry, working in collaborative groups, multisensory strategies, integration of literacy, appropriate use of digital technologies, and making connections to students' lives.

FOSS is designed to make active learning in science engaging for teachers as well as for students. It includes these supports for teachers:

- Complete equipment kits with durable, well-designed materials for all students.
- Detailed *Investigations Guide* with science background for the teacher and focus questions to guide instructional practice and student thinking.
- Multiple strategies for formative assessment.
- Benchmark assessments with scoring guides.
- Strategies for use of science notebooks for novice and experienced users.
- *FOSS Science Resources*, a book of course-specific readings.
- The FOSS website with course-integrated online activities for use in school or at home, suggested extension activities, and extensive online support for teachers.

FOSS Program Goals

FOSS Program Goals

Systemic Reform

FOSS provides schools and school systems with a program that addresses the community science-achievement standards. The FOSS Program prepares students by helping them acquire the knowledge and thinking capacity appropriate for world citizens.

The FOSS Program design makes it appropriate for reform efforts on all scales. It reflects the core ideas to be incorporated into the next-generation science standards. It meets with the approval of science and technology companies working in collaboration with school systems, and it has demonstrated its effectiveness with diverse student and teacher populations in major urban reform efforts. The use of science notebooks and formative-assessment strategies in FOSS redefines the role of science in a school—the way that teachers engage in science teaching with one another as professionals and with students as learners, and the way that students engage in science learning with the teacher and with one another. FOSS takes students and teachers beyond the classroom walls to establish larger communities of learners.

BRIDGING RESEARCH INTO PRACTICE

The FOSS Program is built on the assumptions that understanding core scientific knowledge and how science functions is essential for citizenship, that all teachers can teach science, and that all students can learn science. The guiding principles of the FOSS design, described below, are derived from research and confirmed through FOSS developers' extensive experience with teachers and students in typical American classrooms.

Understanding of science develops over time. FOSS has elaborated learning or content progressions for core ideas in science for kindergarten through grade 8. Developing the learning progressions involves identifying successively more sophisticated ways of thinking about core ideas over multiple years. "If mastery of a core idea in a science discipline is the ultimate educational destination, then well-designed learning progressions provide a map of the routes that can be taken to reach that destination" (National Research Council, *A Framework for K–12 Science Education*, 2011).

Focusing on a limited number of topics in science avoids shallow coverage and provides more time to explore core science ideas in depth. Research emphasizes that fewer topics experienced in greater depth produces much better learning than many topics briefly visited. FOSS affirms this research. FOSS courses provide long-term engagement (10–12 weeks) with important science ideas. Furthermore, courses build upon one another within and across each strand, progressively moving students toward the grand ideas of science. The core ideas of science are difficult and complex, never learned in one lesson or in one class year.

FOSS Next Generation—K–8 Sequence

		PHYSICAL SCIENCE		EARTH SCIENCE		LIFE SCIENCE	
		MATTER	ENERGY AND CHANGE	ATMOSPHERE AND EARTH	ROCKS AND LANDFORMS	STRUCTURE/ FUNCTION	COMPLEX SYSTEMS
8	6–8	Waves; Gravity and Kinetic Energy; Chemical Interactions; Electromagnetic Force; Variables and Design		Planetary Science; Earth History; Weather and Water		Heredity and Adaptation; Human Systems Interactions; Populations and Ecosystems; Diversity of Life	
	5	Mixtures and Solutions		Earth and Sun		Living Systems	
	4		Energy		Soils, Rocks, and Landforms	Environments	
	3	Motion and Matter		Water and Climate		Structures of Life	
	2	Solids and Liquids			Pebbles, Sand, and Silt	Insects and Plants	
	1		Sound and Light	Air and Weather		Plants and Animals	
K	K	Materials and Motion		Trees and Weather		Animals Two by Two	

FOSS Program Goals

A5

FOSS Program Goals

Science is more than a body of knowledge. How well you think is often more important than how much you know. In addition to the science content framework, every FOSS course provides opportunities for students to engage in and understand science practices, and many courses explore issues related to engineering practices and the use of natural resources. FOSS uses these science and engineering practices.

- Asking questions (for science) and defining problems (for engineering)
- Developing and using models
- Planning and carrying out investigations
- Analyzing and interpreting data
- Using mathematics, information and computer technology, and computational thinking
- Constructing explanations (for science) and designing solutions (for engineering)
- Engaging in argument from evidence
- Obtaining, evaluating, and communicating information

Science is inherently interesting, and children are natural investigators. It is widely accepted that children learn science concepts best by doing science. Doing science means hands-on experiences with objects, organisms, and systems. Hands-on activities are motivating for students, and they stimulate inquiry and curiosity. For these reasons, FOSS is committed to providing the best possible materials and the most effective procedures for deeply engaging students with scientific concepts. FOSS students at all grade levels investigate, experiment, gather data, organize results, and draw conclusions based on their own actions. The information gathered in such activities enhances the development of science and engineering practices.

Education is an adventure in self-discovery. Science provides the opportunity to connect to students' interests and experiences. Prior experiences and individual learning styles are important considerations for developing understanding. Observing is often equated with seeing, but in the FOSS Program all senses are used to promote greater understanding. FOSS evolved from pioneering work done in the 1970s with students with disabilities. The legacy of that work is that FOSS investigations naturally use multisensory methods to accommodate students with physical and learning disabilities and also to maximize information gathering for all students. A number of tools, such as the FOSS syringe and balance, were originally designed to serve the needs of students with disabilities.

Formative assessment is a powerful tool to promote learning and can change the culture of the learning environment. Formative assessment in FOSS creates a community of reflective practice. Teachers and students make up the community and establish norms of mutual support, trust, respect, and collaboration. The goal of the community is that everyone will demonstrate progress and will learn and grow.

Science-centered language development promotes learning in all areas. Effective use of science notebooks can promote reflective thinking and contribute to lifelong learning. Research has shown that when language-arts experiences are embedded within the context of learning science, students improve in their ability to use their language skills. Students are motivated to read to find out information, and to share their experiences both verbally and in writing.

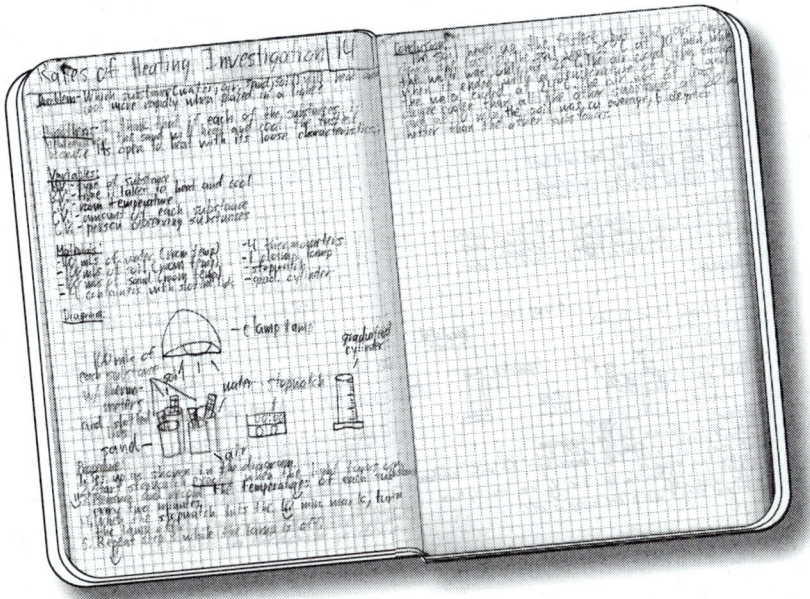

Experiences out of the classroom develop awareness of community. By extending classroom learning into the local region and community, FOSS brings the science concepts and principles to life. In the process of extending classroom learning to the natural world and utilizing community resources, students will develop a relationship with learning that extends beyond the classroom walls.

FOSS Program Goals

A7

FOSS Program Goals

FOSS NEXT GENERATION K–8 SCOPE AND SEQUENCE

Grade	Physical Science	Earth Science	Life Science
6–8	Waves* / Gravity and Kinetic Energy*	Planetary Science	Heredity and Adaptation* / Human Systems Interactions*
	Chemical Interactions	Earth History	Populations and Ecosystems
	Electromagnetic Force* / Variables and Design*	Weather and Water	Diversity of Life
5	Mixtures and Solutions	Earth and Sun	Living Systems
4	Energy	Soils, Rocks, and Landforms	Environments
3	Motion and Matter	Water and Climate	Structures of Life
2	Solids and Liquids	Pebbles, Sand, and Silt	Insects and Plants
1	Sound and Light	Air and Weather	Plants and Animals
K	Materials and Motion	Trees and Weather	Animals Two by Two

*Half-length course

FOSS is a research-based science curriculum for grades K–8 developed at the Lawrence Hall of Science, University of California, Berkeley. FOSS is also an ongoing research project dedicated to improving the learning and teaching of science. The FOSS project began over 25 years ago during a time of growing concern that our nation was not providing young students with an adequate science education. The FOSS Program materials are designed to meet the challenge of providing meaningful science education for all students in diverse American classrooms and to prepare them for life in the 21st century. Development of the FOSS Program was, and continues to be, guided by advances in the understanding of how people think and learn.

With the initial support of the National Science Foundation and continued support from the University of California, Berkeley, and School Specialty, Inc., the FOSS Program has evolved into a curriculum for all students and their teachers, grades K–8. The current editions of FOSS are the result of a rich collaboration among the FOSS/Lawrence Hall of Science development staff; the FOSS product development team at School Specialty; assessment specialists, educational researchers, and scientists; and dedicated professionals in the classroom and their students, administrators, and families.

We acknowledge the thousands of FOSS educators who have embraced the notion that science is an active process, and we thank them for their significant contributions to the development and implementation of the FOSS Program.

Science Notebooks in Middle School

Science Notebooks in Middle School

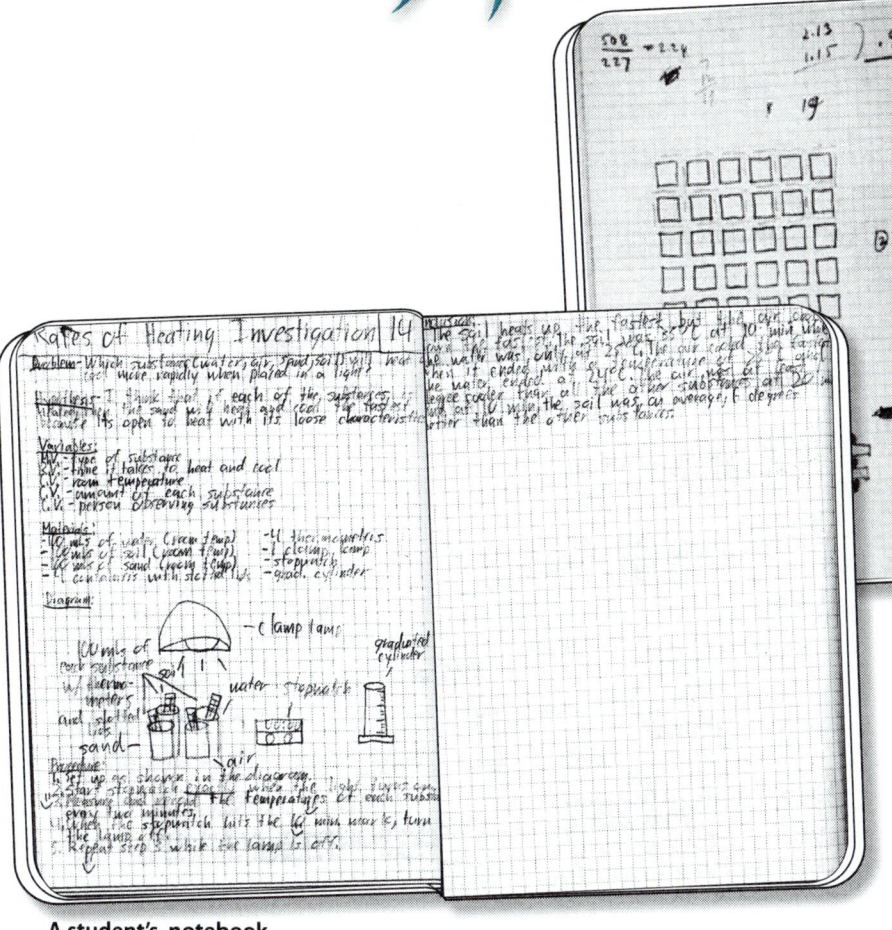

A scientist's notebook

A student's notebook

INTRODUCTION

Scientists keep notebooks. The scientist's notebook is a detailed record of his or her engagement with scientific phenomena. It is a personal representation of experiences, observations, and thinking—an integral part of the process of doing scientific work. A scientist's notebook is a continuously updated history of the development of scientific knowledge and reasoning. The notebook organizes the huge body of knowledge and makes it easier for a scientist to work. As developing scientists, FOSS students are encouraged to incorporate notebooks into their science learning. First and foremost, the notebook is a tool for student learning.

Contents

Introduction	B1
Notebook Benefits	B2
Getting Started	B5
Notebook Components	B12
Focusing the Investigation	B14
Data Acquisition and Organization	B16
Making Sense of Data	B18
Next-Step Strategies	B22
Using Notebooks to Improve Student Learning	B25
Derivative Products	B28

Science Notebooks in Middle School

NOTEBOOK BENEFITS

Engaging in active science is one part experience and two parts making sense of the experience. Science notebooks help students with the sense-making part by providing two major benefits: documentation and cognitive engagement.

Benefits to Students

Science notebooks centralize students' data. When data are displayed in functional ways, students can think about the data more effectively. A well-kept notebook is a useful reference document. When students have forgotten a fact or relationship that they learned earlier in their studies, they can look it up. Learning to reference previous discoveries and knowledge structures is important.

Documentation: an organized record. As students become more accomplished at keeping notebooks, their work will become better organized and efficient. Tables, graphs, charts, drawings, and labeled illustrations will become standard means for representing and displaying data. A complete and accurate record of learning allows students to reconstruct the sequence of learning events and relive the experience. Discussions about science among students, students and teachers, or students, teachers, and families, have more meaning when they are supported by authentic documentation in students' notebooks. Questions and ideas generated by experimentation or discussion can be recorded for future investigation.

From the Human Brain and Senses Course

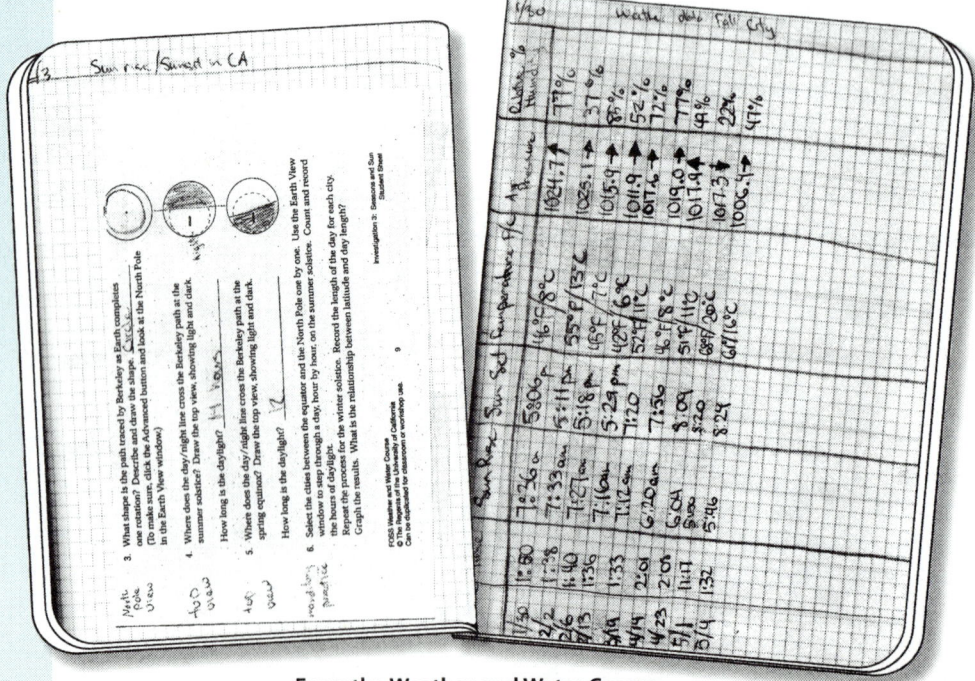

From the Weather and Water Course

B2 — Full Option Science System

Cognitive engagement. Once data are recorded and organized in an efficient manner in science notebooks, students can think about the data and draw conclusions about the way the world works. Their data are the raw materials that students use to forge concepts and relationships from their experiences and observations.

Writing stimulates active reasoning. There is a direct relationship between the formation of concepts and the rigors of expressing them in words. Writing requires students to impose discipline on their thoughts. When you ask students to generate derivative products (summary reports, detailed explanations, posters, oral presentations, etc.) as evidence of learning, the process will be much more efficient and meaningful because they have a coherent, detailed notebook for reference.

When students use notebooks as an integral part of their science studies, they think critically about their thinking. This reflective thinking can be encouraged by notebook entries that present opportunities for self-assessment. Self-assessment motivates students to rethink and restate their scientific understanding. Revising their notebook entries helps students clarify their understanding of the science concepts under investigation. By writing explanations, students clarify what they know and expose what they don't know.

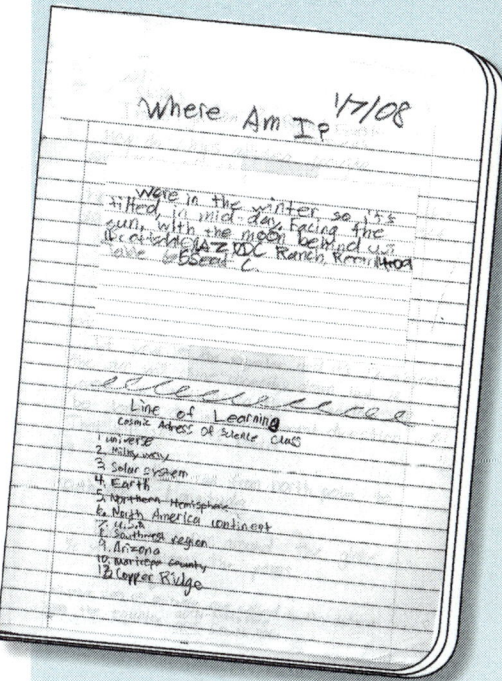

From the Planetary Science Course

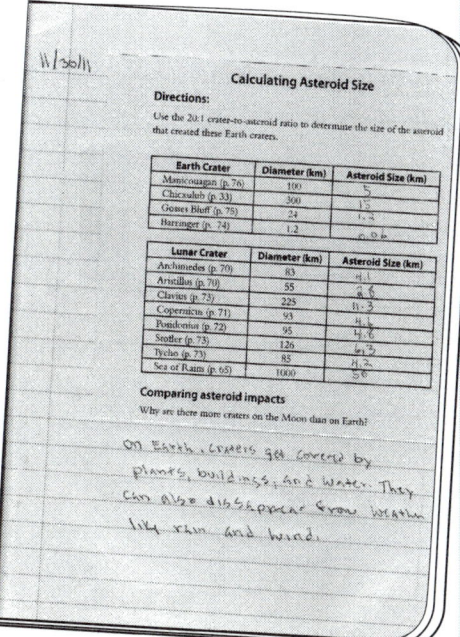

From the Planetary Science Course

Science Notebooks in Middle School

Science Notebooks in Middle School

Benefits to Teachers

In FOSS, the unit of instruction is the course—a sequence of conceptually related learning experiences that leads to a set of learning outcomes. A science notebook helps you think about and communicate the conceptual structure of the course you are teaching.

Assessment. From the assessment point of view, a science notebook is a collection of student-generated artifacts that exhibit learning. You can informally assess student skills, such as using charts to record data, in real time while students are working with materials. At other times, you might collect student work samples and review them for insights or errors in conceptual understanding. This valuable information helps you plan the next steps of instruction. Students' data analysis, sense making, and reflection provide a measure of the quality and quantity of student learning. The notebook itself should not be graded, though certain assignments might be graded and placed in the notebook.

Medium for feedback. The science notebook provides an excellent medium for providing feedback to individual students regarding their work. Productive feedback calls for students to read a teacher comment, think about the issue it raises, and act on it. The comment may ask for clarification, an example, additional information, precise vocabulary, or a review of previous work in the notebook. In this way, you can determine whether a problem with the student work relates to a flawed understanding of the science content or a breakdown in communication skills.

Focus for professional discussions. The student notebook also acts as a focal point for discussion about student learning at several levels. First, a student's work can be the subject of a conversation between you and the student. By acting as a critical mentor, you can call attention to ways a student can improve the notebook, and help him or her learn how to use the notebook as a reference. You can also review and discuss the science notebook during family conferences. Science notebooks shared among teachers in a study group or other professional-development environment can effectively demonstrate recording techniques, individual styles, various levels of work quality, and so on. Just as students can learn notebook strategies from one another, teachers can learn notebook skills from one another.

GETTING STARTED

A middle school science notebook is more than just a collection of science work, notes, field-trip permission slips, and all the other types of documents that tend to accumulate in a student's three-ring binder or backpack. By organizing the science work systematically into a bound composition book, students create a thematic record of their experiences, thoughts, plans, reflections, and questions as they work through a topic in science.

The science notebook is more than just formal lab reports; it is a record of a student's entire journey through a progression of science concepts. Where elementary school students typically need additional help structuring and organizing their written work, middle school students should be encouraged to develop their organizational skills and take some ownership in creating deliberate records of their science learning, even though they may still require some pointers and specific scaffolding from you.

In addition, the science notebook provides a personal space where students can explore their understanding of science concepts by writing down ideas and being allowed to "mess around" with their thinking. Students are encouraged to look back on their ideas throughout the course to self-assess their conceptual development and record new thoughts. With this purpose of the science notebook in mind, you may need to refine your own thinking around what should or should not be included as a part of the science notebook, as well as expectations about grading and analyzing student work.

Science Notebooks in Middle School

Rules of Engagement

Teachers and students should be clear about the conventions students will honor in their notebook entries. Typically, the rules of grammar and spelling are fairly relaxed so as not to inhibit the flow of expression during notebook entries. This also helps students develop a sense of ownership in their notebooks, a place where they are free to write in their own style. When students generate derivative products using information in the notebooks, such as reports, you might require students to exercise more rigorous language-arts conventions.

In addition to written entries, students should be encouraged to use a wide range of other means for recording and communicating, including charts, tables, graphs, drawings, graphics, color codes, numbers, and artifacts attached to the notebook pages. By expanding the options for making notebook entries, each student will find his or her most efficient, expressive way to capture and organize information for later retrieval.

Enhanced Classroom Discussion

One of the benefits of using notebooks is that you will elicit responses to key discussion questions from all students, not just the handful of verbally enthusiastic students in the class. When you ask students to write down their thoughts after you pose a question, all students have time to engage deeply with the question and organize their thoughts. When you ask students to share their answers, those who needed more time to process the question and organize their thinking will be ready to verbalize their responses and become involved in a class discussion.

When students can use their notebooks as a reference during the ensuing discussion, they won't feel put on the spot. At some points, you might ask students to share only what they wrote in their notebooks, to remind them to focus their thoughts while writing. As the class shares ideas during discussions, students can add new ideas to their notebooks under a line of learning (see next-step strategies). Even if some students are still reticent, having students write after a question is posed prevents them from automatically disengaging from conversations.

Notebook Structure

FOSS recommends that students keep their notebooks in 8" × 10" bound composition books. At the most advanced level, students are responsible for creating the entire science notebook from blank pages in their composition books. Experienced students determine when to use their notebooks, how to organize space, what methods of documentation to use, and how to flag important information. This level of notebook use will not be realized quickly; it will likely require systematic development by an entire teaching staff over time.

At the beginning, notebook practice is often highly structured, using prepared sheets from the FOSS notebook masters. You can photocopy and distribute these sheets to students as needed during the investigations. Sheets are sized to fit in a standard composition book. Students glue or tape the sheets into their notebooks. This allows some flexibility between glued-in notebook sheets and blank pages where students can do additional writing, drawings, and other documentation. Prepared notebook sheets are helpful organizers for students with challenges such as learning disabilities or with developing English skills. This model is the most efficient means for obtaining the most productive work from inexperienced middle school students.

From the Planetary Science Course

Science Notebooks in Middle School

Science Notebooks in Middle School

To make it easy for new FOSS teachers to implement a beginning student notebook, Delta Education sells copies of the printed *FOSS Science Notebook* in English for all FOSS middle school courses. Electronic versions of the notebook sheets can be downloaded free of charge at www.FOSSweb.com.

Each *FOSS Science Notebook* is a bound set of the notebook sheets for the course plus extra blank sheets throughout the notebook for students to write focus or inquiry questions, record and organize data, make sense of their thinking, and write summaries. There are also blank pages at the end to develop an index of science vocabulary.

The questions, statements, and graphic organizers on the notebook sheets provide guidance for students and scaffolding for teachers. When the notebook sheets are organized as a series, they constitute a highly structured precursor to an autonomously generated science notebook.

Developing Notebook Skills

Students will initially need more guidance from you. You will need to describe what and when to record, and to model organizational techniques. As the year advances, the notebook work will become increasingly student centered. As the body of work in the notebook grows, students will have more and more examples of useful techniques for reference. This self-sufficiency reduces the amount of guidance you need to provide, and reinforces students' appreciation of their own record of learning.

This gradual shift toward student-centered use of the notebook applies to any number of notebook skills, including developing headers for each page (day, time, date, title, etc.); using space efficiently on the page; preparing graphs, graphic organizers, and labeled illustrations; and attaching artifacts (sand samples, dried flowers, photographs, etc.). For instance, when students first display their data in a two-coordinate graph, the graph might be completely set up for them, so that they simply plot the data. As the year progresses, they will be expected to produce graphs with less and less support, until they are doing so without any assistance from you.

Science Notebooks in Middle School

Organizing Science Notebooks

Four organizational components of the notebook should be planned right from the outset: a table of contents, page numbering, entry format, and an index.

Table of contents. Students should reserve the first three to five pages of their notebook for the table of contents. They will add to it systematically as they proceed through the course. The table of contents should include the date, title, and page number for each entry. The title could be based on the names of the investigations in the course, the specific activities undertaken, the concepts learned, a focus question for each investigation, or some other schema that makes sense to everyone.

Page numbering. Each page should have a number. These are referenced in the table of contents as the notebook progresses.

Entry format. During each class session, students will document their learning. Certain information will appear in every record, such as the date and title. Other forms of documentation will vary, including different types of written entries and artifacts, such as a multimedia printout. Some teachers ask their students to start each new entry at the top of the next available page. Others simply leave a modest space before a new entry. Sometimes it is necessary to leave space for work that will be completed on a separate piece of paper and glued or taped in later. Students might also leave space after a response, so that they can add to it at a later time.

Index. Scientific academic language is important. FOSS strives to have students use precise, accurate vocabulary at all times in their writing and conversations. To help them learn scientific vocabulary, students should set up an index at the end of their notebooks. It is not usually possible for students to enter the words in alphabetical order, as they will be acquired as the course advances. Instead, students could use several pages at the end of the notebook blocked out in 24 squares, and assign one or more letters to each square. Students write the new vocabulary word or phrase in the appropriate square and tag it with the page number of the notebook on which the word is defined. By developing vocabulary in context, students construct meaning through the inquiry process, and by organizing the words in an index, they strengthen their science notebooks as a documentary tool of their science learning. As another alternative, students can also define the word within these squares with the page references.

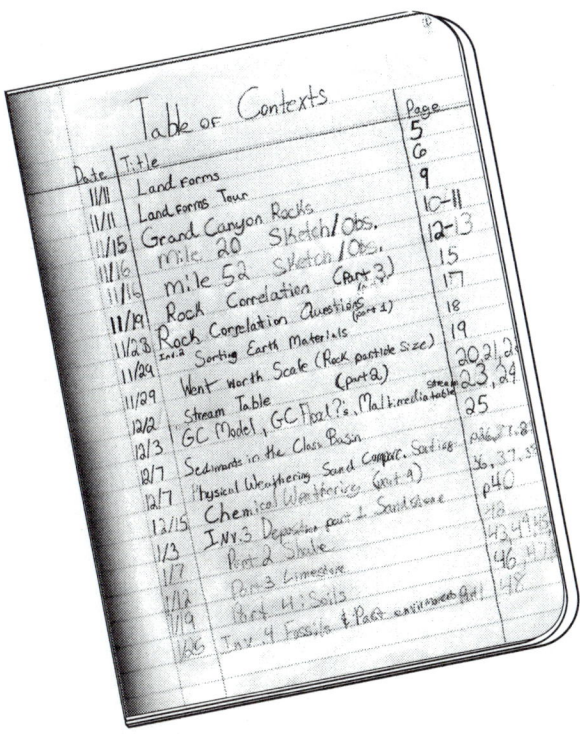

Table of contents

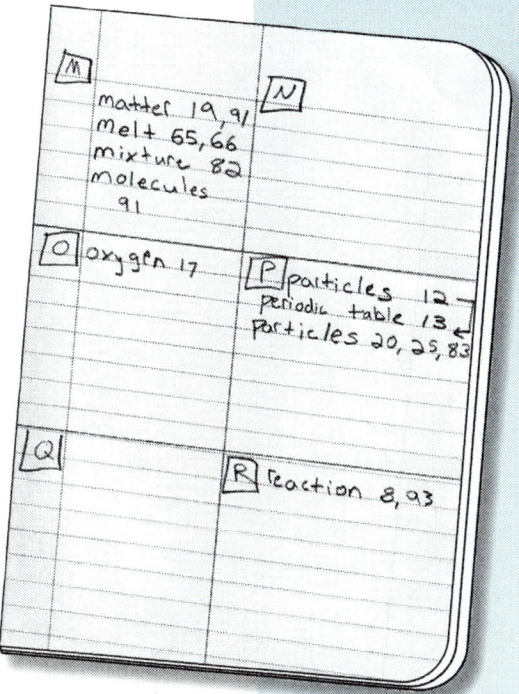

A science notebook index

Science Notebooks in Middle School

Science Notebooks in Middle School

NOTEBOOK COMPONENTS

Four general types of notebook entries, or components, give the science notebook conceptual shape and direction. These structures don't prescribe a step-by-step procedure for how to prepare the notebook, but they do provide some overall guidance. The general arc of an investigation starts with a question or challenge, proceeds with an activity and data acquisition, continues to sense making, and ends with next steps such as reflection and self-assessment.

All four components are not necessary during each class session, but over the course of an investigation, each component will be visited at least once. It may be useful to keep these four components in mind as you systematically guide students through their notebook entries. The components are overviewed here and described in greater detail on the following pages.

Focusing the investigation. Each part of each FOSS investigation includes a focus question, which students transcribe into their notebooks. Focus questions are embedded in the teacher step-by-step instructions and explicitly labeled. The focus question establishes the direction and conceptual challenge for that part of the investigation. For instance, when students investigate the origins of sand and sandstone in the **Earth History Course**, they start by writing,

➤ *Which came first, sand or sandstone?*

The question focuses both students and you on the learning goals for the activity. Students may start by formulating a plan, formally or informally, for answering the focus question. The goal of the plan is to obtain a satisfactory answer to the focus question, which will be revisited and answered later in the investigation.

Data acquisition and organization. After students have established a plan, they collect data. Students can acquire data from carefully planned experiments, accurate measurements, systematic observations, free explorations, or accidental discoveries. It doesn't matter what process produces the data; the critically important point is that students obtain data and record it. It may be necessary to reorganize and display the data for efficient analysis, often by organizing a data table. The data display is key to making sense of the science inquiry.

Making sense of data. Once students have collected and displayed their data, they need to analyze it to learn something about the natural world. In this component of the notebook, students write explanatory statements that answer the focus question. You can formalize this component by asking students to use an established protocol such as a sentence starter, or the explanation can be purely a thoughtful effort by each student. Explanations may be incorrect or incomplete at this point, but students can remedy this during the final notebook entry, when they have an opportunity to continue processing what they've learned. Unfortunately, this piece is often forgotten in the classroom during the rush to finish the lesson and move on. But without sense making and reflection (the final phase of science inquiry), students might see the lesson as a fun activity without connecting the experience to the big ideas that are being developed in the course.

Next-step strategies. The final component of an investigation brings students back to their notebooks by engaging in a next-step strategy, such as reflection and self-assessment, that moves their understanding forward. This component is the capstone on a purposeful series of experiences designed to guide students to understand the concept originally presented in the focus question. After making sense of the data, and making new claims about the topic at hand, students should go back to their earlier thinking and note their changing ideas and new findings. This reflective process helps students cement their new ideas.

A student organizes and makes sense of data in the Chemical Interactions Course.

Science Notebooks in Middle School

B13

Science Notebooks in Middle School

Focusing the Investigation

Focus question. The first notebook entry in most investigations is the focus question. Focus questions are embedded in the teacher step-by-step instructions and explicitly labeled. You can write the question on the board or project it for students to transcribe into their notebooks. The focus question serves to focus students and you on the inquiry for the day. It is not always answered immediately, but rather hangs in the air while the investigation goes forward. Students always revisit their initial responses later in the investigation.

Quick write. A quick write (or quick draw) can be used in addition to a focus question. Quick writes can be completed on a quarter sheet of paper or an index card so you can collect, review, and return them to students to be taped or glued into their notebooks and used for self-assessment later in the investigation.

In the **Diversity of Life Course**, you ask,

> ➤ *What is life?*

For a quick write, students write an answer immediately, before instruction occurs. The quick write provides insight into what students think about certain phenomena before you begin instruction. When responding to the question, students should be encouraged to write down their thoughts, even if they don't feel confident in knowing the answer.

Knowing students' preconceptions will help you know what concepts need the most attention during the investigation. Make sure students date their entries for later reference. Read through students' writing and tally the important points to focus on. Quick writes should not be graded.

Planning. After students enter the focus question or complete a quick write in their notebooks, they plan their investigation. (In some investigations, planning is irrelevant to the task at hand.) Planning may be detailed or intuitive, formal or informal, depending on the requirement of the investigation. Plans might include lists (including materials, things to remember), step-by-step procedures, and experimental design. Some FOSS notebook masters guide students through a planning process specific to the task at hand.

Lists. Science notebooks often include lists of things to think about, materials to get, or words to remember. A materials list is a good organizer that helps students anticipate actions they will take. A list of variables to be controlled clarifies the purpose of an experiment. Simple lists of dates for observations or of the people responsible for completing a task may be useful.

Step-by-step procedures. Middle school students need to develop skills for writing sequential procedures. For example, in the **Chemical Interactions Course**, students write a procedure to answer these questions.

> ➤ *Is there a limit to the amount of substance that will dissolve in a certain amount of liquid?*
>
> ➤ *If so, is the amount that will dissolve the same for all substances?*

Students need to recall what they know about the materials, develop a procedure for accurately measuring the amount of a substance that is added to the water, and agree on a definition of "saturated." To check a procedure for errors or omissions, students can trade notebooks and attempt to follow another student's instructions to complete the task.

Experimental design. Some work with materials requires a structured experimental plan. In the **Planetary Science Course**, students pursue this focus question.

> ➤ *Are Moon craters the result of volcanoes or impacts?*

Students plan an experiment to determine what affects the size and shape of craters on the Moon. They use information they gathered during the open exploration of craters made in flour to develop a strategy for evaluating the effect of changing the variable of a projectile's height or mass. Each lab group agrees on which variable they will change and then designs a sound experimental procedure that they can refer to during the active investigation.

Science Notebooks in Middle School

Data Acquisition and Organization

Because observation is the starting point for answering the focus question, data records should be

- clearly related to the focus question;
- accurate and precise;
- organized for efficient reference.

Data handling can have two subcomponents: data acquisition and data display. Data acquisition is making and recording observations (measurements). The data record can be composed of words, phrases, numbers, and drawings. Data display reorganizes the data in a logical way to facilitate thinking. The display can be a graph, chart, calendar, or other graphic organizer.

Early in a student's experience with notebooks, the record may be disorganized and incomplete, and the display may need guidance. The FOSS notebook masters are designed to help students with data collection and organization. You may initially introduce conventional data-display methods, such as those found in the FOSS notebook masters, but soon students will need opportunities to independently select appropriate data displays. As students become more familiar with collecting and organizing data, you might have them create their own records. With practice, students will become skilled at determining what form of recording to use in various situations, and how best to display the data for analysis.

Narratives. For most students, the most intuitive approach to recording data is narrative—using words, sentence fragments, and numbers in a more or less sequential manner. As students make a new observation, they record it below the previous entry, followed by the next observation, and so on. Some observations, such as a record of weather changes in the **Weather and Water Course** or the interactions of organisms in miniecosystems in the **Populations and Ecosystems Course**, are appropriately recorded in narrative form.

Drawings. A picture is worth a thousand words, and a labeled picture is even more useful. When students use a microscope to discover cells in the *Elodea* leaf and observe and draw structures of microorganisms in the **Diversity of Life Course**, a labeled illustration is the most efficient way to record data.

Charts and tables. An efficient way to record many kinds of data is a chart or table. How do you introduce this skill into the shared knowledge of the classroom? One way is to call for attention during an investigation and demonstrate how to perform the operation. Or you can let students record the data as they like, and observe their methods. There may be one or more groups that invent an appropriate table. During processing time, ask this group to share its method with the class. If no group has spontaneously produced an effective table, you might challenge the class to come up with an easier way to display the data, and turn the skill-development introduction into a problem-solving session.

With experience, students will recognize when a table or chart is appropriate for recording data. When students make similar observations on a series of objects, such as rock samples in the **Earth History Course**, a table with columns is an efficient way to organize observations for easy comparison.

Artifacts. Occasionally an investigation will produce two-dimensional artifacts that students can tape or glue directly into a science notebook. The mounted flower parts in the **Diversity of Life Course** and the sand samples card from the **Earth History Course** can become a permanent part of the record of learning.

Graphs and graphic tools. Reorganizing data into logical, easy-to-use graphic tools is typically necessary for data analysis. Graphs allow easy comparison (bar graph), quick statistical analysis of frequency data (histogram or line plot), and visual confirmation of a relationship between variables (two-coordinate graph). The **Variables and Design Course** offers many opportunities for students to collect data and organize the data into graphs. Students collect data from air trolleys traveling at different speeds, graph the data, and use the resulting graphs to understand how slope of a motion graph can indicate speed. Other graphic tools, such as Venn diagrams, pie charts, and concept maps, help students make connections.

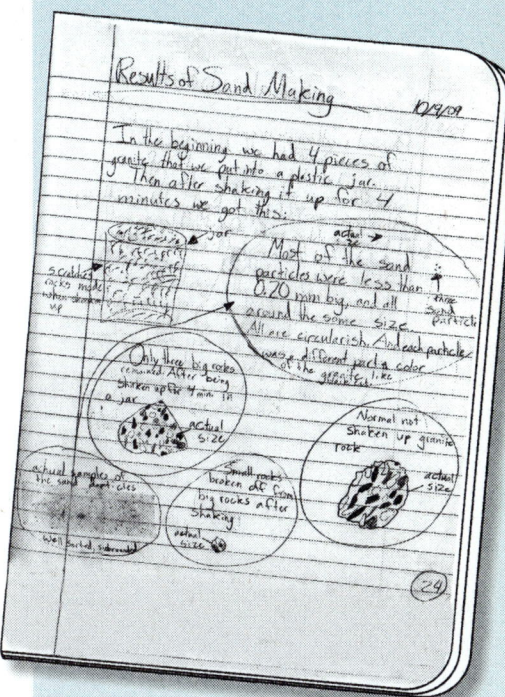

Drawing and artifact from the Earth History Course

Science Notebooks in Middle School

B17

Science Notebooks in Middle School

Making Sense of Data

After collecting and organizing data, the student's next task is to answer the focus question. Students can generate an explanation as an unassisted narrative, but in many instances you might need to use supports such as the FOSS notebook masters to guide the development of a coherent and complete response to the question. Several other support structures for sense making are described below.

Development of vocabulary. Vocabulary is better introduced after students have experienced the new word(s) in context. This sequence provides a cognitive basis for students to connect accurate and precise language to their real-life experiences. Lists of new vocabulary words in the index reinforce new words and organize them for easy reference.

Data analysis. Interpreting data requires the ability to look for patterns, trends, outliers, and potential causes. Students should be encouraged to develop a habit of looking for patterns and relationships within the data collected. Frequently, this is accomplished by creating a graph with numerical data. In the **Populations and Ecosystems Course**, students review field data acquired by ecologists at Mono Lake to determine how biotic and abiotic factors affect the populations of organisms found in the lake.

Graphic organizers. Students can benefit from organizers that help them look at similarities and differences. A compare-and-contrast chart can help students make a transition from their collected data and experiences to making and writing comparisons. It is sometimes easier for students to use than a Venn diagram, and is commonly referred to as a box-and-T chart (as popularized in *Writing in Science: How to Scaffold Instruction to Support Learning*, listed in the Bibliography section).

In this strategy, students draw a box at the top of the notebook page and label it "similar" or "same." On the bottom of the notebook page, they draw a *T*. At the top of each wing of the *T*, they label the objects being compared. Students look at their data, use the *T* to identify differences for each item, and use the "similar" box to list all the characteristics that the two objects have in common. For example, a box-and-T chart comparing characteristics of extrusive and intrusive igneous rocks in the **Earth History Course** might look like this.

similar

extrusive	intrusive

Students can use the completed box-and-T chart to begin writing comparisons. It is usually easier for students to complete their chart on a separate piece of paper, so they can fill it in as they refer to their data. They affix the completed chart into their notebooks after they have made their comparisons.

Claims and evidence. A claim is an assertion about how the natural world works. Claims should always be supported by evidence—statements that are directly correlated with data. The evidence should refer to specific observations, relationships that are displayed in graphs, tables of data that show trends or patterns, dates, measurements, and so on. A claims-and-evidence construction is a sophisticated, rich display of student learning and thinking. It also shows how the data students collected is directly connected to what they learned.

Science Notebooks in Middle School

Frames and prompts. One way to get students to organize their thinking is by providing sentence frames for them to complete.

- *I used to think _____, but now I think _____.*
- *The most important thing to remember about Moon phases is _____.*
- *One new thing I learned about adaptation is _____.*

Prompts also direct students to the content they should be thinking about, but provide more latitude for generating responses. For students who are learning English or who struggle with writing, assistive structures like sentence frames can help them communicate their thinking while they learn the nuances of science writing. The prompts used most often in the FOSS notebook masters take the form of questions for students to answer. In the **Weather and Water Course**, students answer the quick-write question

➤ *What causes seasons?*

After modeling an Earth/Sun system and reviewing solar angle and solar concentration, students revisit their quick write to revise and expand on their original explanations.

- *I used to think seasons were caused by _____, but now I know _____.*

Careful prompts scaffold students by helping them communicate their thinking but do not do the thinking for them. As students progress in communication ability, you might provide frames less frequently.

Conclusions and predictions. At the end of an investigation (major conceptual sequence), it might be appropriate for students to write a summary to succinctly communicate what they have learned. This is where students can make predictions based on their understanding of a principle or relationship. For instance, after completing the investigation of condensation and dew point in the **Weather and Water Course**, a student might predict the altitude at which clouds would form, based on weather-balloon data. Or, after examining ecosystem interactions between biotic and abiotic factors in the **Populations and Ecosystems Course**, students will predict how various human interactions could affect the ecosystem. The conclusion or prediction will frequently indicate the degree to which a student can apply new knowledge to real-world situations. A prediction can also be the springboard for further inquiry.

Full Option Science System

Generating new questions. Does the investigation connect to a student's personal interests? Does the outcome suggest a question or pique a student's curiosity? The science classroom is most exciting when students are generating their own questions for further investigation based on class or personal experiences. The notebook is an excellent place to capture students' musings and record thoughts that might otherwise be lost.

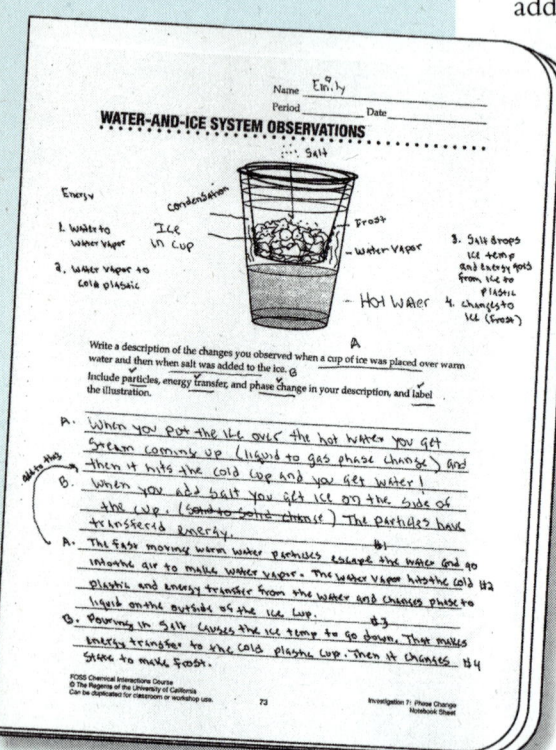

Science Notebooks in Middle School

A student's revised work for the Chemical Interactions Course

Next-Step Strategies

The goal of the FOSS curriculum is for students to develop accurate, durable knowledge of the science content under investigation. Students' initial conceptions are frequently incomplete or confused, requiring additional thought to become fully functional. The science notebook is a useful place to guide reflection and revision. Typically students commit their understanding in writing and reflect in three locations.

- Explanatory narratives in notebooks
- Response sheets incorporated into the notebook
- Written work on I-Checks

These three categories of written work provide information about student learning for you *and* a record of thinking for students that they can reflect on and revise. Scientists constantly refine and clarify their ideas about how the natural world works. They read scientific articles, consult with other scientists, and attend conferences. They incorporate new information into their thinking about the subject they are researching. This reflective process can result in deeper understanding or a complete revision of thinking.

After completing one of the expositions of knowledge—a written conclusion, response sheet, or benchmark assessment—students should receive additional instruction or information via a next-step strategy. They will use this information later to complete self-assessment by reviewing their original written work, making judgments about its accuracy and completeness, and writing a revised explanation. You can use any of a number of techniques for providing the additional information to students.

- Group compare-and-share discussion
- Think/pair/share reading
- Whole-class critique of an explanation by an anonymous student
- Identifying key points for a class list
- Whole-class discussion of a presentation by one student

After one of the information-generating processes, students compare the "best answer" to their own answer and rework their explanations if they can no longer defend their original thinking. The revised statement of the science content can take one of several forms.

B22   Full Option Science System

Students might literally revise the original writing, crossing out extraneous or incorrect bits, inserting new or improved information, and completing the passage. At other times, students might reflect on their original work and, after drawing and dating a line of learning (see below), might redraft their explanation from scratch, producing their best explanation of the concept.

During these self-assessment processes, students have to think actively about every aspect of their understanding of the concept and organize their thoughts into a coherent, logical narrative. The learning that takes place during this process is powerful. The relationships between the several elements of the concept become unified and clarified.

The notebook is the best tool for students when preparing for benchmark assessment, such as an I-Check or posttest. Students don't necessarily have the study skills needed to prepare on their own, but using teacher-guided tasks such as key points and traffic lights will turn the preparation process into a valuable exercise. These same strategies can be used after a benchmark assessment when you identify further areas of confusion or misconceptions you want to address with students. Here are four helpful next-step, or self-assessment, strategies.

Line of learning. One technique many teachers find useful in the reflective process is the line of learning. After students have conducted an investigation and entered their initial explanations, they draw and date a line under their original work. As students share ideas and refine their thinking during class discussion, additional experimentation, reading, and teacher feedback, encourage them to make new entries under the line of learning, adding to or revising their original thinking. If the concept is elusive or complex, a second line of learning, followed by more processing and revising, may be appropriate.

The line of learning is a reminder to students that learning is an ongoing process with imperfect products. It points out places in that process where a student made a stride toward full understanding. And the psychological security provided by the line of learning reminds students that they can always draw another line of learning and revise their thinking again. The ability to look back in the science notebook and see concrete evidence of learning gives students confidence and helps them become critical observers of their own learning.

A line of learning used with the Planetary Science Course

Science Notebooks in Middle School

Science Notebooks in Middle School

Traffic lights. In the traffic-lights strategy, students use color to self-assess and indicate how well they understand a concept that they are learning. Green means that the student feels that he or she has a good understanding of the concept. Yellow means that the student is still a bit unsure about his or her understanding. Red means that the student needs help; he or she has little or no understanding of the concept. Students can use colored pencils, markers, colored dots, or colored cards to indicate their understanding. They can mark their own work and then indicate their level of understanding by a show of hands or by holding up colored cards. This strategy gives students practice in self-assessment and helps you monitor students' current understanding. You should follow up by looking at student work to ensure that they actually do understand the content that they marked with green.

Three C's. Another approach to revision is to apply the three C's—confirm, correct, complete—to the original work. Students indicate ideas that were correct with a number or a color, code statements needing correction with a second number or color, and assign a third number or color to give additional information that completes the entry.

Key points. Students do not necessarily connect the investigative experience with the key concepts and processes taught in the lesson. It is essential to give students an opportunity to reflect on their experiences and find meaning in those experiences. They should be challenged to use their experiences and data to either confirm or reject their current understanding of the natural world. As students form supportable ideas about a concept, those ideas should be noted as key points, posted in the room, and written in their notebooks. New evidence, to support or clarify an idea, can be added to the chart as the course progresses. If an idea doesn't hold up under further investigation, a line can be drawn through the key point to indicate a change in thinking. A key-points activity is embedded near the end of each investigation to help students organize their thinking and prepare for benchmark assessment.

USING NOTEBOOKS TO IMPROVE STUDENT LEARNING

Notebook entries should not be graded. Research has shown that more learning occurs when students get only comments on written work in their notebooks, not grades or a combination of comments and grades.

If your school district requires a certain number of grades each week, select certain work products that you want to grade and have students turn in that work separate from the notebook. After grading, return the piece to students to insert into their notebooks, so that all their work is in one place.

It may be difficult to stop using grades or a rubric for notebook assessment. But providing feedback that moves learning forward, however difficult, has benefits that make it worth the effort. The key to using written feedback for formative assessment is to make feedback timely and specific, and to provide time for students to act on the feedback by revising or correcting work right in their own notebook.

Teacher Feedback

Student written work often exposes weaknesses in understanding—or so it appears. It is important for you to find out if the flaw results from poor understanding of the science or from imprecise communication. You can use the notebook to provide two types of feedback to the student: to ask for clarification or additional information, and to ask probing questions that will help students move forward in their thinking. Respecting the student's space is important, so rather than writing directly in the notebook, attach a self-stick note, which can be removed after the student has taken appropriate action.

The most effective forms of feedback relate to the content of the work. Here are some examples.

> ➤ You wrote that seasons are caused by Earth's tilt. Does Earth's tilt change during its orbit?
>
> ➤ What evidence can you use to support your claim that Moon craters are caused by impacts? Hint: Think of our experiments in class.

Nonspecific feedback, such as stars, pluses, smiley faces, and "good job!", or ambiguous critiques, such as "try again," "put more thought into this," and "not enough," are less effective and should not be used. Feedback that guides students to think about the content of their work and gives suggestions for how to improve are productive instructional strategies.

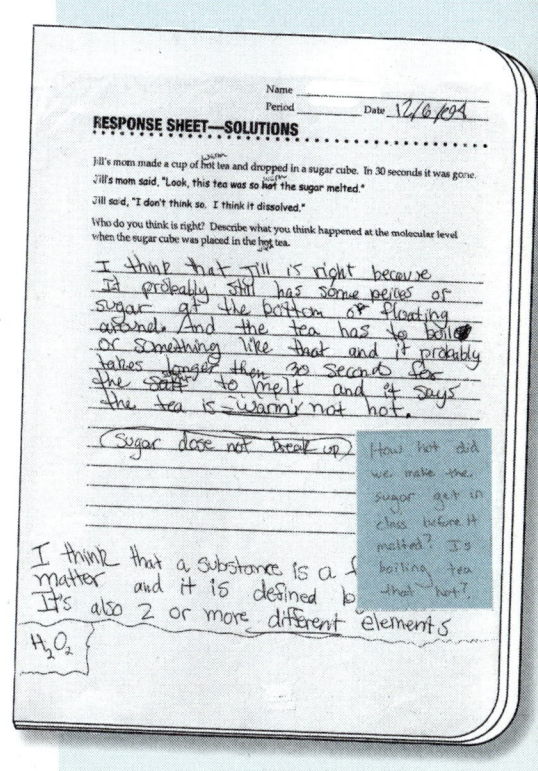

Feedback given during the Chemical Interactions Course

Science Notebooks in Middle School

Science Notebooks in Middle School

Here are some appropriate generic feedback questions to write or use verbally while you circulate in the class.

- ➤ *What vocabulary have you learned that will help you describe _____ ?*
- ➤ *Can you include an example from class to support your ideas?*
- ➤ *Include more detail about _____ .*
- ➤ *Check your data to make sure this is accurate.*
- ➤ *What do you mean by _____ ?*
- ➤ *When you record your data, what unit should you use?*

When students return to their notebooks and respond to the feedback, you will have additional information to help you discriminate between learning and communication difficulties. Another critical component of teacher feedback is providing comments to students in a timely manner, so that they can review their work before engaging in benchmark assessment or moving on to other big ideas in the course.

In middle school, you face the challenge of having a large number of students. This may mean collecting a portion of students' notebooks on alternate days. Set a specific focus for your feedback, such as a data table or conclusion, so you aren't trying to look at everything every time.

To help students improve their writing, you might have individuals share notebook entries aloud in their collaborative groups, followed by feedback from a partner or the group. This valuable tool must be very structured to create a safe environment, including ground rules about acceptable feedback and comments.

A good way to develop these skills is to model constructive feedback with the class, using a student-work sample from a notebook. Use a sample from a previous year with the name and any identifying characteristics removed. Project it for the class to practice giving feedback.

Formative Assessment

With students recording more of their thinking in an organized notebook, you have a tool to better understand the progress of students and any misconceptions that are typically not revealed until the benchmark assessment. One way to monitor student progress is during class while they are responding to a prompt. Circulate from group to group, and read notebook entries over students' shoulders. This is a good time to have short conversations with individuals or small groups to gain information about the level of student understanding. Take care to respect the privacy of students who are not comfortable sharing their work during the writing process.

If you want to look at work that is already completed in the notebook, ask students to open their notebooks to the page that you want to review and put them in a designated location. Or consider having students complete the work on a separate piece of paper or an index card. Students can leave a blank page in their notebooks, or label it with a header as a placeholder, until they get the work back and tape it or glue it in place. This makes looking at student work much easier, and the record of learning that the student is creating in the notebook remains intact.

When time is limited, you might select a sample of students from each class, alternating the sample group each time, to get a representative sample of student thinking. This is particularly useful following a quick write.

Once you have some information about student thinking, you can make teaching decisions about moving ahead to a benchmark assessment, going back to a previous concept, or spending more time making sense of a concept. Benchmark assessments can also be used as formative assessment. You might choose to administer an I-Check, score the assessment to find problem areas, and then revisit critical concepts before moving on to the next investigation. Students can use reflection and self-assessment techniques to revisit and build on their original exam responses.

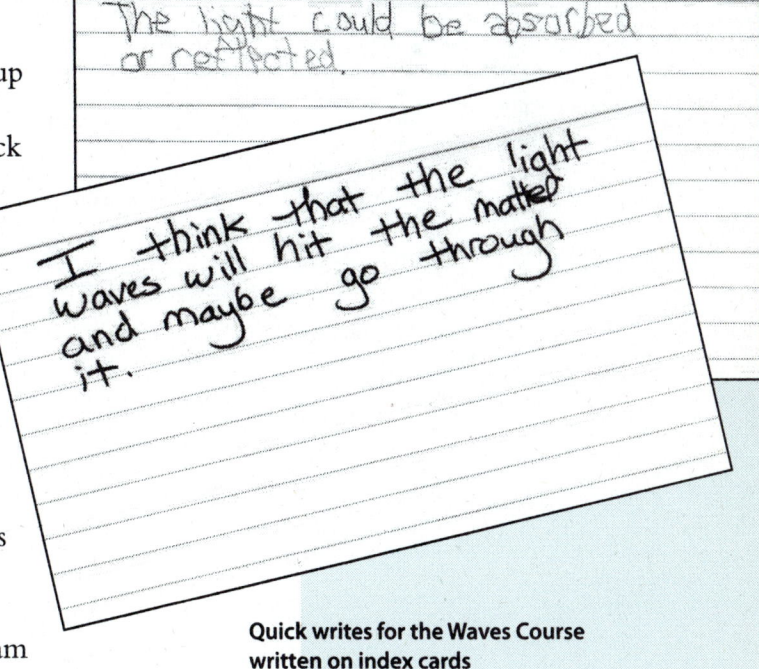

Quick writes for the Waves Course written on index cards

Science Notebooks in Middle School

DERIVATIVE PRODUCTS

On occasion, you might ask students to produce science projects of various kinds: summary reports, detailed explanations, end-of-course projects, oral reports, or posters. Students should use their notebooks as a reference when developing their reports. You could ask them to make a checklist of science concepts and pieces of evidence, with specific page references, extracted from their notebooks. They can then use this checklist to ensure that all important points have been included in the derivative work.

The process of developing a project has feedback benefits, too. While students are developing projects using their notebooks, they have the opportunity to self-monitor the organization and content of the notebook. This offers valuable feedback on locating and extracting useful information. You might want to discuss possible changes students would make next time they start a new science notebook.

Homework is another form of derivative product, as it is an extension of the experimentation started in class. Carefully selected homework assignments enhance students' science learning. Homework suggestions and/or extension activities are included at the end of each investigation. For example, in the **Heredity and Adaptation Course**, after using an online activity in class to predict genetic variation, students are asked to complete a follow-up online simulation as homework. In the **Electromagnetic Force Course**, after students test properties of magnets in class, they are asked to look for examples of magnets in household objects outside of the classroom.

Science-Centered Language Development in Middle School

Science-Centered Language Development in Middle School

Contents

Introduction C1

The Role of Language in Scientific and Engineering Practices C3

Speaking and Listening Domain C6

Writing Domain C12

Reading Domain C18

Science-Vocabulary Development C26

English-Language Development C31

References C43

Reading and writing are inextricably linked to the very nature and fabric of science, and, by extension, to learning science.

Stephen P. Norris and Linda M. Phillips, "How Literacy in Its Fundamental Sense Is Central to Scientific Literacy"

INTRODUCTION

In this chapter, we explore the ways reading, writing, speaking, and listening are interwoven in effective science instruction at the secondary level. To engage fully in the enterprise of science and engineering, students must record and communicate observations and explanations, and read about and discuss the discoveries and ideas of others. This becomes increasingly challenging at the secondary level. Texts become more complex; writing requires fluency of academic language, including domain-specific vocabulary. Here we identify strategies that support sense making. The active investigations, student science notebooks, *FOSS Science Resources* readings, multimedia, and formative assessments provide rich contexts in which students develop and exercise thinking processes and communication skills. Students develop scientific literacy through experiences with the natural world around them in real and authentic ways and use language to inquire, process information, and communicate their thinking about the objects, organisms, and phenomena they are studying. We refer to the acquisition and building of language skills necessary for scientific literacy as science-centered language development.

Science-Centered Language Development in Middle School

Language plays two crucial roles in science learning: (1) it facilitates the communication of conceptual and procedural knowledge, questions, and propositions (external; public), and (2) it mediates thinking, a process necessary for understanding (internal; private). These are also the ways scientists use language: to communicate with one another about their inquiries, procedures, and understandings; to transform their observations into ideas; and to create meaning and new ideas from their work and the work of others. For students, language development is intimately involved in their learning about the natural world. Active-learning science provides a real and engaging context for developing literacy; language-arts skills and strategies support conceptual development and scientific and engineering practices. For example, the skills and strategies used for reading comprehension, writing expository text, and oral discourse are applied when students are recording their observations, making sense of science content, and communicating their ideas. Students' use of language improves when they discuss, write, and read about the concepts explored in each investigation.

We begin our exploration of science and language by focusing on language functions and how specific language functions are used in science to facilitate information acquisition and processing (thinking). Then we address issues related to the specific language domains—speaking and listening, writing, and reading. Each section addresses

- how skills in that domain are developed and exercised in FOSS science investigations;
- literacy strategies that are integrated purposefully into the FOSS investigations; and
- suggestions for additional literacy strategies that both enhance student learning in science and develop or exercise academic-language skills.

Following the domain discussions is a section on science-vocabulary development, with scaffolding strategies for supporting all learners. The last section covers language-development strategies specifically for English learners.

▶ **NOTE**
The term *English learners* refers to students who are learning to understand English. This includes students who speak English as a second language and native English speakers who need additional support to use language effectively.

THE ROLE OF LANGUAGE IN SCIENTIFIC AND ENGINEERING PRACTICES

Language functions are the purpose for which speech or writing is used and involve both vocabulary and grammatical structure. Understanding and using language functions appropriately is important in effective communication. Students use numerous language functions in all disciplines to mediate communication and facilitate thinking (e.g., they plan, compare, discuss, apply, design, draw, and provide evidence).

In science, language functions facilitate scientific and engineering practices. For example, when students are *collecting data*, they are using language functions to identify, label, enumerate, compare, estimate, and measure. When students are *constructing explanations*, they are using language functions to analyze, communicate, discuss, evaluate, and justify.

A Framework for K–12 Science Education (National Research Council 2012) states that "Students cannot comprehend scientific practices, nor fully appreciate the nature of scientific knowledge itself, without directly experiencing the practices for themselves." Each of these scientific and engineering practices uses multiple language functions. Often, these language functions are part of an internal dialogue weighing the merits of various explanations—what we call thinking. The more language functions with which we are facile, the more effective and creative our thinking can be.

The scientific and engineering practices are listed below, along with a sample of the language functions that are exercised when students are effectively engaged in that practice. (Practices are bold; language functions are italic.)

Asking questions and defining problems

- *Ask* questions about objects, organisms, systems, and events in the natural and human-made world (science).
- *Ask* questions to *define* and *clarify* a problem, *determine criteria* for solutions, and *identify* constraints (engineering).

Planning and carrying out investigations

- *Plan* and conduct investigations in the laboratory and in the field to gather appropriate data (*describe* procedures, *determine* observations to *record*, *decide* which variables to control) or to gather data essential for *specifying* and *testing* engineering designs.

Examples of Language Functions

Analyze
Apply
Ask
Clarify
Classify
Communicate
Compare
Conclude
Construct
Critique
Describe
Design
Develop
Discuss
Distinguish
Draw
Enumerate
Estimate
Evaluate
Experiment
Explain
Formulate
Generalize
Group
Identify
Infer
Interpret
Justify
Label
List
Make a claim
Measure
Model
Observe
Organize
Plan
Predict
Provide evidence
Reason
Record
Represent
Revise
Sequence
Solve
Sort
Strategize
Summarize
Synthesize

Science-Centered Language Development in Middle School

Science-Centered Language Development in Middle School

Analyzing and interpreting data
- Use a range of tools (numbers, words, tables, graphs, images, diagrams, equations) to *organize* observations (data) in order to *identify* significant features and patterns.

Developing and using models
- Use models to help *develop explanations, make predictions*, and *analyze* existing systems, and *recognize* strengths and limitations of the models.

Using mathematics and computational thinking
- Use mathematics and computation to *represent* physical variables and their relationships.

Constructing explanations and designing solutions
- *Construct* logical explanations of phenomena, or *propose solutions* that incorporate current understanding or a model that represents it and is consistent with the available evidence.

Engaging in argument from evidence
- *Defend* explanations, *formulate evidence* based on data, *examine* one's own understanding in light of evidence offered by others, and collaborate with peers in searching for explanations.

Obtaining, evaluating, and communicating information
- *Communicate* ideas and the results of inquiry—orally and in writing—with tables, diagrams, graphs, and equations and in *discussion* with peers.

Research supports the claim that when students are intentionally using language functions in thinking about and communicating in science, they improve not only science content knowledge, but also language-arts and mathematics skills (Ostlund, 1998; Lieberman and Hoody, 1998). Language functions play a central role in science as a key cognitive tool for developing higher-order thinking and problem-solving abilities that, in turn, support academic literacy in all subject areas.

Here is an example of how an experienced teacher can provide an opportunity for students to exercise language functions in FOSS. In the **Planetary Science Course**, one piece of content we expect students to have acquired by the end of the course is

- The lower the angle at which light strikes a surface, the lower the density of the light energy.

The scientific practices the teacher wants the class to focus on are *developing and using models* and *constructing explanations*.

The language functions students will exercise while engaging in these scientific practices include *comparing, modeling, analyzing,* and *explaining*. The teacher understands that these language functions are appropriate to the purpose of the science investigation and support the *Common Core State Standards for English Language Arts and Literacy in Science* (CCSS), in which grades 6–8 students will "write arguments focused on discipline-specific content . . . support claim[s] with logical reasoning and relevant, accurate data and evidence that demonstrate an understanding of the topic" (National Governors Association Center for Best Practices, Council of Chief State School Officers, 2010).

- Students will *compare* the area covered by the same beam of light (from a flashlight) at different angles to *explain* the relationship between the angle and density of light energy.

> **▶ CCSS NOTE**
> This example supports
> CCSS.ELA-Literacy.WHST.6–8.1.b.

The teacher can support the use of language functions by providing structures such as sentence frames.

- As _____, then _____.

  As the angle increases, then the light beam becomes smaller and more circular.

- When I changed _____, then _____.

  When I changed the angle of the light beam, then the concentration of light hitting the floor changed.

- The greater/smaller _____, the _____.

  The smaller the angle of the light beam, the more the light beam spread out.

- I think _____, because _____.

  I think the smaller spot of light receives more energy than the larger spot because the light concentration is greatest when light shines directly down on a surface and there is no beam spreading.

Science-Centered Language Development in Middle School

Science-Centered Language Development in Middle School

SPEAKING AND LISTENING DOMAIN

The FOSS investigations are designed to engage students in productive oral discourse. Talking requires students to process and organize what they are learning. Listening to and evaluating peers' ideas calls on students to apply their knowledge and to sharpen their reasoning skills. Guiding students in instructive discussions is critical to the development of conceptual understanding of the science content and the ability to think and reason scientifically. It also addresses a key middle school CCSS Speaking and Listening anchor standard that students "engage effectively in a range of collaborative discussions (one-on-one, in groups, and teacher-led) with diverse partners on [grade-level] topics, texts, and issues, building on others' ideas and expressing their own clearly."

FOSS investigations start with a discussion—a review to activate prior knowledge, presentation of a focus question, or a challenge to motivate and engage active thinking. During the active investigation, students talk with one another in small groups, share their observations and discoveries, point out connections, ask questions, and start to build explanations. The discussion icon in the sidebar of the *Investigations Guide* indicates when small-group discussions should take place.

Throughout the activity, the *Investigations Guide* indicates where it is appropriate to pause for whole-class discussions to guide conceptual understanding. The *Investigations Guide* provides you with discussion questions to help stimulate student thinking and support sense making. At times, it may be beneficial to use sentence frames or standard prompts to scaffold the use of effective language functions and structures. Allowing students a few minutes to write in their notebooks prior to sharing their answers also helps those who need more time to process and organize their thoughts.

On the following pages are some suggestions for providing structure to those discussions and for scaffolding productive discourse when needed. Using the protocols that follow will ensure inclusion of all students in discussions.

▶ **CCSS NOTE**
This example supports CCSS.ELA-Literacy.SL.6.1, CCSS.ELA-Literacy.SL.7.1, and CCSS.ELA-Literacy.SL.8.1.

▶ **NOTE**
Additional notebook strategies can be found in the Science Notebooks in Middle School chapter in *Teacher Resources* and online at www.FOSSweb.com.

Partner and Small-Group Discussion Protocols

Whenever possible, give students time to talk with a partner or in a small group before conducting a whole-class discussion. This provides all students with a chance to formulate their thinking, express their ideas, practice using the appropriate science vocabulary, and receive input from peers. Listening to others communicate different ways of thinking about the same information from a variety of perspectives helps students negotiate the difficult path of sense making for themselves.

Dyads. Students pair up and take turns either answering a question or expressing an idea. Each student has 1 minute to talk while the other student listens. While student A is talking, student B practices attentive listening. Student B makes eye contact with student A, but cannot respond verbally. After 1 minute, the roles reverse.

Here's an example from the **Chemical Interactions Course**. After reviewing the results of eight reactions recorded in their notebooks, you ask students to pair up and take turns sharing which two substances they think constitute the mystery mixture and their reasons for selecting those two. The language objective is for students to compare their test results and make inferences based on their observations and what they know about chemical reactions (orally and in writing). These sentence frames can be written on the board to scaffold student thinking and conversation.

- I think the two substances in the mystery mixture are _____ and _____.

- My evidence is _____.

Partner parade. Students form two lines facing each other. Present a question, an idea, an object, or an image as a prompt for students to discuss. Give them 1 minute to greet the person in front of them and discuss the prompt. After 1 minute, call time. Have the first student in one of the lines move to the end of the line, and have the rest of the students in that line shift one step sideways so that everyone has a new partner. (Students in the other line do not move.) Give students a new prompt to discuss for 1 minute with their new partners. This can also be done by having students form two concentric circles. After each prompt, the inner circle rotates.

For example, when students are just beginning the **Earth History Course** investigation on igneous rock, you may want to assess prior knowledge about Earth's layers. Give each student a picture from an assortment of related images such as volcanoes, magma, a diagram of Earth's layers, crystals, and so forth, and have students line up facing

Partner and Small-Group Discussion Protocols
- *Dyads*
- *Partner parade*
- *Put in your two cents*

Science-Centered Language Development in Middle School

Science-Centered Language Development in Middle School

each other in two lines or in concentric circles. For the first round, ask, "What do you observe in the image on your card?" For the second round, ask, "What can you infer from the image?" For the third round, ask, "What questions do you have about the image?" The language objective is for students to describe their observations, infer how the landform formed, and reflect upon and relate any experiences they may have had with a similar landform. These sentence frames can be used to scaffold student discussion.

- I observe _____, _____, and _____.
- I think this shows _____ because _____.
- I wonder _____.

Put in your two cents. For small-group discussions, give each student two pennies or similar objects to use as talking tokens. Each student takes a turn putting a penny in the center of the table and sharing his or her idea. Once all have shared, each student takes a turn putting in the other penny and responding to what others in the group have said. For example,

- I agree (or don't agree) with _____ because _____.

Here's an example from the **Diversity of Life Course**. In their notebooks, students have recorded the amount of water lost from their vials containing celery with and without leaves. They discover a discrepancy in the amount of water lost and the mass of the celery. Where did the water go? Students are struggling to form an explanation. The language objective is for students to compare their results and infer that there is a relationship between the amount of water lost and the number of leaves the celery has. You give each student two pennies, and in groups of four, they take turns putting in their two cents. For the first round, each student answers the question "Where did the water go?" They use the frame

- I think the water _____.
- My evidence is _____.

On the second round, each student states whether he or she agrees or disagrees with someone else in the group and why, using the sentence frame.

Whole-Class Discussion Supports

The whole-class discussion is a critical part of sense making. After students have had the active learning experience and have talked with their peers in pairs and/or small groups, sharing their observations with the whole class sets the stage for developing conventional explanatory models. Discrepant events, differing results, and other surprises are discussed, analyzed, and resolved. It is important that students realize that science is a process of finding out about the world around them. This is done through asking questions, testing ideas, forming explanations, and subjecting those explanations to logical scrutiny, that is, argumentation. Leading students through productive discussion helps them connect their observations and the abstract symbols (words) that represent and explain those observations. Whole-class discussion also provides an opportunity for you to interject an accurate and precise verbal summary as a model of the kind of thinking you are seeking. Facilitating effective whole-class discussions takes skill, practice, a shared set of norms, and patience. In the long run, students will have a better grasp of the content and will improve their ability to think independently and communicate effectively.

Norms should be established so that students know what is expected during science discussions.

- Science content and practices are the focus.
- Everyone participates (speaking and listening).
- Ideas and experiences are shared, accepted, and valued. Everyone is respectful of one another.
- Claims are supported by evidence.
- Challenges (debate and argument) are part of the quest for complete understanding.

A variety of whole-class discussion techniques can be introduced and practiced during science instruction that address the CCSS Speaking and Listening standards for students to "present claims and findings [e.g., argument, narrative, informative, summary presentations], emphasizing salient points in a focused, coherent manner with relevant evidence, sound valid reasoning, and well-chosen details; use appropriate eye contact, adequate volume, and clear pronunciation."

For example, during science talk, students are reminded to practice attentive listening, stay focused on the speaker, ask questions, and respond appropriately. In addition, in order for students to develop and practice their reasoning skills, they need to know the language forms

Whole-Class Discussion Supports
- *Sentence frames*
- *Guiding questions*

TEACHING NOTE
Let students know that scientists change their minds based on new evidence. It is expected that students will revise their thinking, based on evidence presented in discussions.

▶ CCSS NOTE
This example supports CCSS.ELA-Literacy.SL.6.4, CCSS.ELA-Literacy.SL.7.4, and CCSS.ELA-Literacy.SL.8.4, (grade 8 quoted here).

Science-Centered Language Development in Middle School

Science-Centered Language Development in Middle School

TEACHING NOTE

Encourage "science talk." Allow time for students to engage in discussions that build on other students' observations and reasoning. After an investigation, use a teacher- or student-generated question, and either just listen or facilitate the interaction with questions to encourage expression of ideas among students.

and structures and the behaviors used in evidence-based debate and argument, such as using data to support claims, disagreeing respectfully, and asking probing questions (Winokur and Worth, 2006).

Explicitly model the language structures appropriate for active discussions, and encourage students to use them when responding to guiding questions and during science talks.

Sentence frames. These samples can be posted as a scaffold as students develop their reasoning and oral participation skills.

- I think _____, because _____.
- I predict _____, because _____.
- I claim _____; my evidence is _____.
- I agree with _____ that _____.
- My idea is similar/related to _____'s idea.
- I learned/discovered/heard that _____.
- <Name> explained _____ to me.
- <Name> shared _____ with me.
- We decided/agreed that _____.
- Our group sees it differently, because _____.
- We have different observations/results. Some of us found that _____. One group member thinks that _____.
- We had a different approach/idea/solution/answer: _____.

Guiding questions. The Investigations Guide provides questions to help concentrate student thinking on the concepts introduced in the investigation. Guiding questions should be used during the whole-class discussion to facilitate sense making. Here are some other open-ended questions that help guide student thinking and promote discussion.

- What did you notice when _____?
- What do you think will happen if _____?
- How might you explain _____? What is your evidence?
- What connections can you make between _____ and _____?

C10

Full Option Science System

Whole-Class Discussion Protocols

The following tried-and-true participation protocols can be used to enhance whole-class discussions. The purpose of these protocols is to increase meaningful participation by giving all students access to the discussion, allowing students time to think (process), and providing a context for motivation and engagement.

Think-pair-share. When asking for a response to a question posed to the class, give students a minute to think silently. Then, have students pair up with a partner to exchange thoughts before you call on a student to share his or her ideas with the whole class.

Pick a stick. Write each student's name on a craft stick, and keep the sticks handy in a cup at the front of the room. When asking for responses, randomly pick a stick, and call on that student to start the discussion. Continue to select sticks as you continue the discussion. Your name can also be on a stick in the cup. To keep students on their toes, put the selected sticks into a smaller cup hidden inside the larger cup out of view of students. That way students think they may be called again.

Whip around. Each student takes a quick turn sharing a thought or reaction. Questions are phrased to elicit quick responses that can be expressed in one to five words (e.g., "Give an example of a stored-energy source." "What does the word *heat* make you think of?").

Commit and toss. Have students write a response to a question or prompt on a loose piece of paper (Keeley, 2008). Next, tell everyone to crumple up the paper into a ball and toss it to another student. Continue tossing for a few minutes, and then call for students to stop, grab a ball, and read the response silently. Responses can then be shared with partners, small groups, or the whole class. This activity allows students to answer anonymously, so they may be willing to share their thinking more openly.

Group posters. Have small groups design and graphically record their investigation data and conclusions on a quickly generated poster to share with the whole class.

Whole-Class Discussion Protocols
- *Think-pair-share*
- *Pick a stick*
- *Whip around*
- *Commit and toss*
- *Group posters*

Cup within a cup
pick-a-stick container

Science-Centered Language Development in Middle School

WRITING DOMAIN

Information processing is enhanced when students engage in informal writing. When allowed to write expressively without fear of being scorned for incorrect spelling or grammar, students are more apt to organize and express their thoughts in different ways that support their own sense making. Writing in science promotes use of science and engineering practices, thereby developing a deeper engagement with the science content. This type of informal writing also provides a springboard for more formal derivative science writing (Keys, 1999).

Science Notebooks

The science notebook is an effective tool for enhancing learning in science and exercising various forms of writing. Notebooks provide opportunities both for expressive writing (students craft explanatory narratives that make sense of their science experiences) and for practicing informal technical writing (students use organizational structures and writing conventions). Students learn to communicate their thinking in an organized fashion while engaging in the cognitive processes required to develop concepts and build explanations. Having this developmental record of learning also provides an authentic means for assessing students' progress in both scientific thinking and communication skills.

Developing Writing for Literacy in Science

Using student science notebooks in science instruction provides opportunities to address the CCSS for Writing in Science. Grades 6–8 students "write routinely over extended time frames (time for research, reflection, and revision) and shorter time frames (a single sitting or a day or two) for a range of tasks, purposes, and audiences." In addition to providing a structure for recording and analyzing data, notebooks serve as a reference tool from which students can draw information in order to produce derivative products, that is, more formal science writing pieces that have a specific purpose and format. CCSS focus on three text types that students should be writing in science: argument, informational/explanatory writing, and narrative writing. These text types are used in science notebooks and can be developed into derivative products such as reports, articles, brochures, poster boards, electronic presentations, letters, and so forth. Following is a description of these three text types and examples that may be used with FOSS investigations to help students build scientific literacy.

▶ **NOTE**
For more information about supporting science-notebook development, see the Science Notebooks in Middle School chapter.

▶ **CCSS NOTE**
This example supports CCSS.ELA-Literacy.W.10.

Engaging in Argument

In science, middle school students make claims in the form of statements or conclusions that answer questions or address problems. CCSS Appendix A describes that for students to use "data in a scientifically acceptable form, students marshal evidence and draw on their understanding of scientific concepts to argue in support of their claims." Applying the literacy skills necessary for this type of writing concurrently supports the development of critical science and engineering practices—most notably, engaging in argument. According to *A Framework for K–12 Science Education*, upon which the Next Generation Science Standards (NGSS) are based, middle school students are expected to construct a convincing argument that supports or refutes claims for explanations about the natural and designed world in these ways.

In FOSS, this type of writing makes students' thinking visible. Both informally in their notebooks and formally on assessments, students use deductive and inductive reasoning to construct and defend their explanations. In this way, students deepen their science understanding and exercise the language functions necessary for higher-level thinking, for example, comparing, synthesizing, evaluating, and justifying. To support students in both oral and written argumentation, use the questions and prompts in the *Investigations Guide* that encourage students to use evidence, models, and theories to support their arguments. In addition, be prepared for those teachable moments that provide the perfect stage for spontaneous scientific debate. Here are some general questions to help students deepen their writing.

- Why do you agree or disagree with _____?
- How would you prove/disprove _____?
- What data did you use to make that conclusion that _____?
- Why was it better that _____?

Here are the ways engaging in written argument are developed in the FOSS investigations and can be extended through formal writing.

Response sheets. The FOSS response sheets give students practice in constructing arguments by providing hypothetical situations where they have to apply what they have learned in order to evaluate a claim. For example, one of the response sheets in the **Planetary Science Course** asks students to respond to three students' explanations for the seasons. Students write a paragraph to each student with the

---

Engaging in Argument
- *Response sheets*
- *Think questions*
- *I-Checks and surveys/posttests*
- *Persuasive writing*

▶ **CCSS NOTE**
This example supports CCSS.ELA-Literacy.W.1.

Science-Centered Language Development in Middle School

purpose of changing his or her thinking. In order to refute each claim, students must evaluate the validity of the statements and construct arguments based on evidence from the data they've collected during the investigations and logical reasoning that supports their explanation for what causes seasons.

Think questions. Interactive reading in *FOSS Science Resources* is another opportunity for students to engage in written argumentation. Articles include questions that support reading comprehension and extend student thinking about the science content. Asking students to make a claim and provide evidence to support it encourages the use of language functions necessary for higher-level thinking such as evaluating, applying, and justifying. For example, in *FOSS Science Resources: Planetary Science*, students are asked to respond to the following question: Why do you think there are so few craters on Earth and so many on the Moon? After discussion with their peers, students can hone their argumentation skills by writing an argument that answers the question and is supported by the evidence in the *FOSS Science Resources* book as well as data recorded from their experience making model craters.

I-Checks and surveys/posttests. Like the FOSS response sheets, some test items assess students' ability to make a claim and provide evidence to support it. One way is to provide students with data and have them make a claim based on that data and evidence from their prior investigations. Their argument should use logical reasoning to support their ideas. For example, in **Planetary Science**, students are shown images taken from two different planets. They are told that one has a thick atmosphere and the other has no atmosphere. They are asked which image they think came from a planet with an atmosphere and why. Using the images, they can see evidence of craters, and they can draw on their own experiences as well as knowledge acquired through other sources to piece together a logical argument.

Persuasive writing. Formal writing gives students the opportunity to summarize, explain, apply, and evaluate what they have learned in science. It also provides a purpose and audience that motivate students to produce higher-level writing products. The objective of persuasive writing is to convince the reader that a stated interpretation of data is worthwhile and meaningful. In addition to supporting claims with evidence and using logical argument, the writer also uses persuasive techniques such as a call to action. Students can use their informal notebook entries to form the basis of formal persuasive writing in a variety of formats, such as essays, letters, editorials, advertisements, award nominations, informational pamphlets, and petitions. Animal habitats, energy use, weather patterns, landforms, and water sources are just a few science topics that can generate questions and issues for persuasive writing.

Here is a sample of writing frames that can be used to introduce and scaffold persuasive writing (modified from Gibbons, 2002).

Title: _____

The topic of this discussion is _____.

My opinion (position, conclusion) is _____.

There are <number> reasons why I believe this to be true.

First, _____.

Second, _____.

Finally, _____.

On the other hand, some people think _____.

I have also heard people say _____.

However, my claim is that _____ because _____.

Science-Centered Language Development in Middle School

Informational/Explanatory Writing
- *Writing frames*
- *Recursive cycle*

▶ **CCSS NOTE**
Designing, recording, and following procedures in FOSS courses supports CCSS.ELA-Literacy.RST.6–8.3.

▶ **CCSS NOTE**
This example supports CCSS.ELA-Literacy.W.2.

Informational/Explanatory Writing

Informational and explanatory writing requires students to examine and convey complex ideas and information clearly and accurately through the effective selection, organization, and analysis of content. In middle school science, this includes writing scientific procedures and experiments. Described in CCSS Appendix A, informational/explanatory writing answers the questions, What type? What are the components? What are the properties, functions, and behaviors? How does it work? What is happening? Why? In FOSS, this type of writing takes place informally in science notebooks, where students are recording their questions, plans, procedures, data, and answers to the focus questions. It also supports sense making as students attempt to convey what they know in response to questions and prompts, using language functions such as identifying, comparing and contrasting, explaining cause-and-effect relationships, and sequencing.

As an extension of the notebook entries, students can apply their content knowledge to publish formal products such as letters, definitions, procedures, newspaper and magazine articles, posters, pamphlets, and research reports. Strategies such as the writing process (plan, draft, edit, revise, and share) and writing frames (modeling and guiding the use of topic sentences, transition and sequencing words, examples, explanations, and conclusions) can be used to scaffold and help students develop proficiency in science writing.

Writing frames. Here are samples of writing frames (modified from Wellington and Osborne, 2001).

Description

Title: _____

(Identify) The part of the _____ I am describing is the _____.

(Describe) It consists of _____.

(Explain) The function of these parts is _____.

(Example) This drawing shows _____.

Explanation

Title: _____

I want to explain why (how) _____.

An important reason for why (how) this happens is that _____.

Another reason is that _____.

I know this because _____.

Recursive cycle. An effective method for extending students' science learning through writing is the recursive cycle of research (Bereiter, 2002). This strategy emphasizes writing as a process for learning, similar to the way students learn during the active science investigations.

1. Decide on a problem or question to write about.
2. Formulate an idea or a conjecture about the problem or question.
3. Identify a remedy or an answer, and develop a coherent discussion.
4. Gather information (from an experiment, science notebooks, *FOSS Science Resources*, FOSSweb multimedia, books, Internet, interviews, videos, etc.).
5. Reevaluate the problem or question based on what has been learned.
6. Revise the idea or conjecture.
7. Make presentations (reports, posters, electronic presentations, etc.).
8. Identify new needs, and make new plans.

This process can continue for as long as new ideas and questions occur, or students can present a final product in any of the suggested formats.

Narrative Writing

Narrative writing conveys an experience to the reader, usually with sensory detail and a sequence of events. In middle school science, students learn the importance of writing narrative descriptions of their procedures with enough detail and precision to allow others to replicate the experiment. Science also provides a broad landscape of stimulating material for stories, songs, biographies, autobiographies, poems, and plays. Students can enrich their science learning by using organisms or objects as characters; describing habitats and environments as settings; and writing scripts portraying various systems, such as weather patterns, states of matter, and the water, rock, or life cycle.

▶ **CCSS NOTE**
This example supports CCSS.ELA-Literacy.W.7.

▶ **NOTE**
Human characteristics should not be given to organisms (anthropomorphism) in science investigations, only in literacy extensions.

▶ **CCSS NOTE**
This example supports CCSS.ELA-Literacy.W.3.

Science-Centered Language Development in Middle School

Science-Centered Language Development in Middle School

READING DOMAIN

Reading is an integral part of science learning. Just as scientists spend a significant amount of their time reading each other's published works, students need to learn to read scientific text—to read effectively for understanding, with a critical focus on the ideas being presented.

The articles in *FOSS Science Resources* facilitate sense making as students make connections to the science concepts introduced and explored during the active investigations. Concept development is most effective when students are allowed to experience organisms, objects, and phenomena firsthand before engaging the concepts in text. The text and illustrations help students make connections between what they have experienced concretely and the abstract ideas that explain their observations.

FOSS Science Resources provides students with clear and coherent explanations, ways of visualizing important information, and different perspectives to examine and question. As students apply these strategies, they are, in effect, using some of the same scientific thinking processes that promote critical thinking and problem solving. In addition, the text provides a level of complexity appropriate for middle schoolers to develop high-level reading comprehension skills. This development requires support and guidance as students grapple with more complex dimensions of language meaning, structure, and conventions. To become proficient readers of scientific and other academic texts, students must be armed with an array of reading comprehension strategies and have ample opportunities to practice and extend their learning by reading texts that offer new language, new knowledge, and new modes of thought.

Oral discourse and writing are critical for reading comprehension and for helping students make sense of the active investigations. Use the suggested prompts, questions, and strategies in the *Investigations Guide* to support comprehension as students read from *FOSS Science Resources*. For most of the investigation parts, the articles are designed to follow the active investigation and are interspersed throughout the course. This allows students to acquire the necessary background knowledge in context through active experience before tackling the wider-ranging content and relationships presented in the text. Breakpoints in the readings are suggested in the *Investigations Guide* to support student conceptual development. Some questions make connections between the reading and the student's class experience. Other questions help the students consider the writer's intent. Additional strategies for reading are derived from the seven essential strategies that readers use to help them understand what they read (Keene and Zimmermann, 2007).

▶ **CCSS NOTE**
The use of *FOSS Science Resources* supports CCSS.ELA-Literacy.RST.6–8.10.

▶ **CCSS NOTE**
Reading breakpoints in the *Investigations Guide* support CCSS.ELA-Literacy.RST.6–8.8.

- Monitor for meaning: discover when you know and when you don't know.

- Use and create schemata: make connections between the novel and the known; activate and apply background knowledge.

- Ask questions: generate questions before, during, and after reading that reach for deeper engagement with the text.

- Determine importance: decide what matters most, what is worth remembering.

- Infer: combine background knowledge with information from the text to predict, conclude, make judgments, and interpret.

- Use sensory and emotional images: create mental images to deepen and stretch meaning.

- Synthesize: create an evolution of meaning by combining understanding with knowledge from other texts/sources.

Reading Comprehension Strategies

Below are some strategies that enhance the reading of expository texts in general and have proven to be particularly helpful in science. Read and analyze the articles beforehand in order to guide students through the text structures and content more effectively.

Build on background knowledge. Activating prior knowledge is critical for helping students make connections between what they already know and new information. Reading comprehension improves when students have the opportunity to think, discuss, and write about what they know about a topic before reading. Review what students learned from the active investigation, provide prompts for making connections, and ask questions to help students recall past experiences and previous exposure to concepts related to the reading.

Create an anticipation guide. Create true-or-false statements related to the key ideas in the reading selection. Ask students to indicate if they agree or disagree with each statement before reading, then have them read the text, looking for the information that supports their true-or-false claims. Anticipation guides connect students to prior knowledge, engage them with the topic, and encourage them to explore their own thinking. To provide a challenge for advanced students, have them come up with the statements for the class.

Draw attention to vocabulary. Check the article for bold faced words students may not know. Review the science words that are already defined in students' notebooks. For new science and nonscience

---

Reading Comprehension Strategies
- *Build on background knowledge*
- *Create an anticipation guide*
- *Draw attention to vocabulary*
- *Preview the text*
- *Turn and talk*
- *Jigsaw text reading*
- *Note making*
- *Summarize and synthesize*
- *3-2-1*
- *Write reflections*
- *Preview and predict*
- *SQ3R*

▶ **CCSS NOTE**
The example of reviewing what students learned from the active investigation supports CCSS.ELA-Literacy.RST.6–8.9.

▶ **CCSS NOTE**
This example supports CCSS.ELA-Literacy.RST.6–8.4.

Science-Centered Language Development in Middle School

Science-Centered Language Development in Middle School

vocabulary words that appear in the reading, have students predict their meanings before reading. During the reading, have students use strategies such as context clues and word structure to see if their predictions were correct. This strategy activates prior knowledge and engages students by encouraging analytical participation with the text.

Preview the text. Give students time to skim through the selection, noting subheads, before reading thoroughly. Point out the particular structure of the text and what discourse markers to look for. For example, most *FOSS Science Resources* articles are written as cause and effect, problem and solution, question and answer, comparison and contrast, description, or sequence. Students will have an easier time making sense of the text if they know what text structure to look for. Model and have students practice analyzing these different types of expository text structures by looking for examples, patterns, and discourse markers. For example, let's look at a passage from *FOSS Science Resources: Planetary Science*.

> An eclipse of the Moon occurs when Earth passes exactly between the Moon and the Sun. [cause and effect] The Moon moves into Earth's shadow during a lunar eclipse. At the time of a full lunar eclipse, Earth's shadow completely covers the disk of the Moon. [description] This is how Earth, the Moon, and the Sun are aligned for a lunar eclipse to be observed. [photograph] Why don't we see a lunar eclipse every month? [question and answer] Because of the tilt of the Moon's orbit around the Earth, Earth's shadow does not fall on the Moon in most months.

Point out how the text in *FOSS Science Resources* is organized (titles, headings, subheadings, questions, and summaries) and if necessary, review how to use the table of contents, glossary, and index. Explain how to scan for formatting features that provide key information (such as boldface type and italics, captions, and framed text) and graphic features (such as tables, graphs, photographs, maps, diagrams, and charts) that help clarify, elaborate, and explain important information in the reading.

While students preview the article, have them focus on the questions that appear in the text, as well as questions at the end of the article. Encourage students to write down questions they have that they think the article will answer.

Turn and talk. When reading as a whole class, stop at key points and have students share their thinking about the selection with the student sitting next to them or with their collaborative group. This strategy helps students process the information and allows everyone to participate in the discussion. When reading in pairs, encourage

---

▶ **NOTE**
Discourse markers are words or phrases that relate one idea to another. Examples are *however, on the other hand,* and *second.*

▶ **CCSS NOTE**
This example supports CCSS.ELA-Literacy.RST.6-8.5 and CCSS.ELA-Literacy.RST.6-8.6.

▶ **CCSS NOTE**
This example supports CCSS.ELA-Literacy.RST.6-8.7.

students to stop and discuss with their partners. One way to encourage engagement and understanding during paired reading is to have students take turns reading aloud a paragraph or section on a certain topic. The one who is listening then summarizes the meaning conveyed in the passage.

Jigsaw text reading. Students work together in small groups (expert teams) to develop a collective understanding of a text. Each expert team is responsible for one portion of the assigned text. The teams read and discuss their portions to gain a solid understanding of the key concepts. They might use graphic organizers to refine and organize the information. Each expert team then presents its piece to the rest of the class. Or form new jigsaw groups that consist of at least one representative from each expert team. Each student shares with the jigsaw group what their team learned from their particular portion of the text. Together, the participants in the jigsaw group fit their individual pieces together to create a complete picture of the content in the article.

Note making. The more students interact with a reading, the better their understanding. Encourage students to become active readers by asking them to make notes as they read. Studies have shown that note making—especially paraphrasing and summarizing—is one of the most effective means for understanding text (Graham and Herbert, 2010; Applebee, 1984). Some investigation parts include notebook sheets that match pages in *FOSS Science Resources*. This allows students to highlight and underline important points, add notes in the margins, and circle words they do not know. Students can also annotate the article by writing thoughts and questions on self-stick notes. Using symbols or codes can help facilitate comprehension monitoring. Here are some possible symbols students can use to communicate their thinking as they interact with text. (Harvey, 1998).

*	interesting
BK	background knowledge
?	question
C	confusing
I	important
L	learning something new
W	wondering
S	surprising

> ▶ **CCSS NOTE**
> This example supports
> CCSS.ELA-Literacy.RST.6–8.10.

Science-Centered Language Development in Middle School

Science-Centered Language Development in Middle School

> **CCSS NOTE**
> This example supports
> CCSS.ELA-Literacy.RST.6–8.1 and
> CCSS.ELA-Literacy.RST.6–8.2.

Students can also use a different set of symbols while making notes about connections: the readings in *FOSS Science Resources* incorporate the active learning that students gain from the investigations, so that they can make authentic text-to-self (T-S) connections. In other words, what they read reminds them of firsthand experiences, making the article more engaging and easier to understand. Text-to-text (T-T) connections are notes students make when they discover a new idea that reminds them of something they've read previously in another text. Text-to-world (T-W) connections involve the text and more global everyday connections to students' lives.

You can model note-making strategies by displaying a selection of text, using a projection system, a document camera, or an interactive whiteboard. As you read the text aloud, model how to write comments on self-stick notes, and use a graphic organizer in a notebook to enhance understanding.

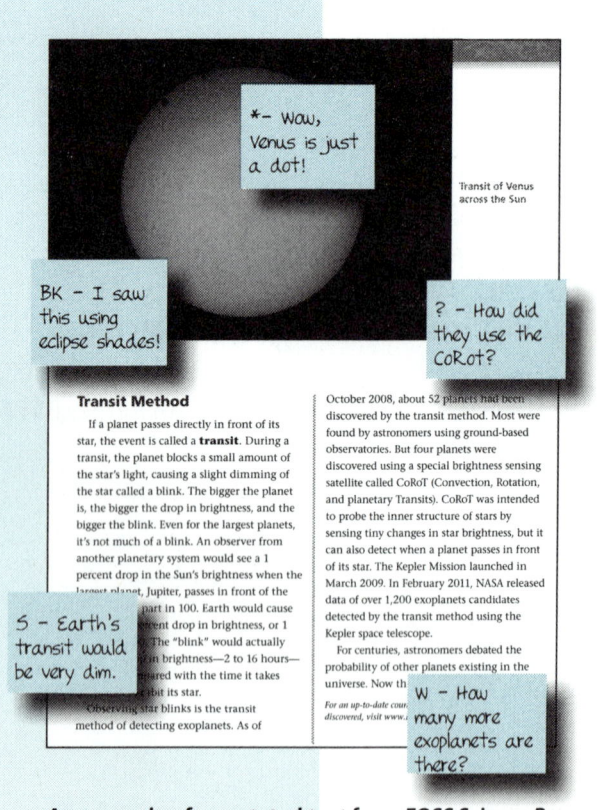

An example of annotated text from *FOSS Science Resources: Planetary Science*

Graphic organizers help students focus on extracting the important information from the reading and analyzing relationships between concepts. This can be done by simply having students make columns in their notebooks to record information and their thinking (Harvey and Goudvis, 2007). Here are two examples.

Notes	Thinking

Facts	Questions	Responses

▶ **CCSS NOTE**
This example supports CCSS.ELA-Literacy.RST.6–8.1 and CCSS.ELA-Literacy.RST.6–8.2.

Summarize and synthesize. Model how to pick out the important parts of the reading selection. Paraphrasing is one way to summarize. Have students write summaries of the reading, using their own words. To scaffold the learning, use graphic organizers to compare and contrast, group, sequence, and show cause and effect. Another method is to have students make two columns in their notebooks. In one column, they record what is important, and in the other, they record their personal responses (what the reading makes them think about). When writing summaries, tell students,

- *Pick out the important ideas.*
- *Restate the main ideas in your own words.*
- *Keep it brief.*

3-2-1. This graphic-organizer strategy gives students the opportunity to synthesize information and formulate questions they still have regarding the concepts covered in an article. In their notebooks, students write three new things they learned, two interesting things worth remembering and sharing, and one question that occurred to them while reading the article. Other options might include three facts, two interesting ideas, and one insight about themselves as learners; three key words, two new ideas, and one thing to think about (modified from Black Hills Special Services Cooperative, 2006).

Write reflections. After reading, ask students to review their notes in their notebooks to make any additions, revisions, or corrections to what they recorded during the reading. This review can be facilitated by using a line of learning. Students draw a line under their original conclusion or under their answer to a question posed at the end of an article. They add any new information as a new narrative entry. The line of learning indicates that what follows represents a change of thinking.

Science-Centered Language Development in Middle School

Preview and predict. Instruct students to independently preview the article, directing attention to the illustrations, photos, boldfaced words, captions, and anything else that draws their attention. Working with a partner, students discuss and write three things they think they will learn from the article. Have partners verbally share their list with another pair of students. The group of four can collaborate to generate one list. Groups report their ideas, and together you create a class list on chart paper.

Read the article aloud, or have students read with a partner aloud or silently. Referring to the preview/prediction list, discuss what students learned. Have them record the most important thing they learned from the reading for comparison with the predictions.

SQ3R. Survey, Question, Read, Recall, Reflect strategy provides an overall structure for before, during, and after reading. Students begin by surveying or previewing the text, looking for features that will help them make predictions about the content. Based on their surveys, students develop questions to answer as they read. They read the selections looking for answers to their questions. Next, they recall what they have learned by retelling a partner and/or recording what they've learned. Finally, they reflect on what they have learned, check to see that they've answered their questions sufficiently, and add any new ideas. Below is a chart students can use to record the SQ3R process in their notebooks.

S Survey	Q Question	R Read	R Recall	R Reflect
Scan the text and record important information.	Ask questions about the subject and what you already know.	Record answers to your questions after you read.	Retell what you learned in your own words.	Did you answer your questions? Record new ideas and comments.

Struggling Readers

For students reading below grade level, the strategies listed on the previous pages can be modified to support reading comprehension by integrating scaffolding strategies such as read-alouds and guided reading. Breaking the reading down into smaller chunks, providing graphic organizers, and modeling reading comprehension strategies can also help students who may be struggling with the text. For additional strategies for English learners, see the supported-reading strategy in the English-Language Development section of this chapter.

Interactive reading aloud. Reading aloud is an effective strategy for enhancing text comprehension. It offers opportunities to model specific reading comprehension strategies and allows students to concentrate on making sense of the content. When modeling, share the thinking processes used to understand the reading (questioning, visualizing, comparing, inferring, summarizing, etc.), then have students share what they observed you thinking about as an active reader.

Guided reading. While the rest of the class is reading independently or in small groups, pull a group aside for a guided reading session. Before reading, review vocabulary words from the investigation and ask questions to activate prior knowledge. Have students preview the text to make predictions, ask questions, and think about text structure. Review reading comprehension strategies they will need to use (monitoring for understanding, asking questions, summarizing, synthesizing, etc.). As students read independently, provide support where needed. Ask questions and provide prompts to guide comprehension. (See the list below for additional strategies.) After reading, have students reflect on what strategies they used to help them understand the text and make connections to the investigation.

- While reading, look for answers to questions and confirm predictions.
- Study graphics, such as pictures, graphs, and tables.
- Reread captions associated with pictures, graphs, and tables.
- Note all italicized and boldfaced words or phrases.
- Reduce reading speed for difficult passages.
- Stop and reread parts that are not clear.
- Read only a section at a time, and summarize after each section.

Struggling Readers
- *Interactive reading aloud*
- *Guided reading*

Science-Centered Language Development in Middle School

SCIENCE-VOCABULARY DEVELOPMENT

Words play two critically important functions in science. First and most important, we play with ideas in our minds, using words. We present ourselves with propositions—possibilities, questions, potential relationships, implications for action, and so on. The process of sorting out these thoughts involves a lot of internal conversation, internal argument, weighing options, and complex linguistic decisions. Once our minds are made up, communicating that decision, conclusion, or explanation in writing or through verbal discourse requires the same command of the vocabulary. Words represent intelligence; acquiring the precise vocabulary and the associated meanings is key to successful scientific thinking and communication.

The words introduced in FOSS investigations represent or relate to fundamental science concepts and should be taught in the context of the investigation. Many of the terms are abstract and are critical to developing science content knowledge and scientific and engineering practices. The goal is for students to use science vocabulary in ways that demonstrate understanding of the concepts the words represent—not to merely recite scripted definitions. The most effective strategies for science-vocabulary development help students make connections to what they already know. These strategies focus on giving new words conceptual meaning through experience; distinguishing between informal, everyday language and academic language; and using the words in meaningful contexts.

Building Conceptual Meaning through Experience

In most instances, students should be presented with new words when they need to know them in the context of the active experience. Words such as *kinetic energy, atmospheric pressure, chemical reaction, photosynthesis*, and *transpiration* are abstract and conceptually loaded. Students will have a much better chance of understanding, assimilating, and remembering the new word (or new meaning) if they can connect it with a concrete experience.

The vocabulary icon appears in the sidebar when students are prompted to record new words in their notebook. The words that appear in bold are critical to understanding the concepts or scientific practices students are learning and applying in the investigation.

When you introduce a new word, students should

- Hear it: students listen as you model the correct contextual use and pronunciation of the word;
- See it: students see the new word written out;
- Say it: students use the new word when discussing their observations and inferences; and
- Write it: students use the new words in context when they write in their notebooks.

Bridging Informal Language to Science Vocabulary

FOSS investigations are designed to tap into students' inquisitive nature and their excitement of discovery in order to encourage lively discussions as they explore materials in creative ways. There should be a lot of talking during the investigations! Your role is to help students connect informal language to the vocabulary used to express specific science concepts. As you circulate during active investigations, you continually model the use of science vocabulary. For example, as students are examining a leaf under the microscope, they will say, "I can see little mouths." You might respond, "Yes, those mouthlike openings are called stomates. They are pores that open and close." Below are some strategies for validating students' conversational language while developing their familiarity with and appreciation for science vocabulary.

Bridging Informal Language to Science Vocabulary
- *Cognitive-content dictionaries*
- *Concept maps*
- *Semantic webs*
- *Word associations*
- *Word sorts*

Cognitive-content dictionaries. Choose a term that is critical for conceptual understanding of the science investigation. Have students write the term, predict its meaning, write the final meaning after class discussion (using primary language or an illustration), and use the term in a sentence.

Cognitive-Content Dictionary	
New term	kinetic energy
Prediction (clues)	something that moves a lot
Final meaning	motion energy
How I would use it in a sentence	Fast-moving particles have more kinetic energy than slow-moving particles.

Science-Centered Language Development in Middle School

Science-Centered Language Development in Middle School

Concept maps. Select six to ten related science words. Have students write them on self-stick notes or cards. Have small groups discuss how the words are related. Students organize words in groups and glue them down or copy them on a sheet of paper. Students draw lines between the related words. On the lines, they write words describing or explaining how the concept words are related.

Semantic webs. Select a vocabulary word, and write it in the center of a piece of paper (or on the board for the whole class). Brainstorm a list of words or ideas that are related to the first word. Group the words and concepts into several categories, and attach them to the central word with lines, forming a web (modified from Hamilton, 2002).

Word associations. In this brainstorming activity, you say a word, and students respond by writing the first word that comes to mind. Then students share their words with the class. This activity builds connections to students' prior frames of reference.

Word sorts. Have students work with a partner to make a set of word cards using new words from the investigation. Have them group the words in different ways, for example, synonyms, root words, and conceptual connections.

Using Science Vocabulary in Context

For a new vocabulary word to become part of a student's functional vocabulary, he or she must have ample opportunities to hear and use it. Vocabulary terms are used in the activities through teacher talk, whole-class and small-group discussions, writing in science notebooks, readings, and assessments. Other methods can also be used to reinforce important vocabulary words and phrases.

Word wall. Use chart paper to record science content and procedural words as they come up during and after the investigations. Students will use this word wall as a reference.

Drawings and diagrams. For English learners and visual learners, use a diagram to review and explain abstract content. Ahead of time, draw an illustration lightly, almost invisibly, with pencil on chart paper. You can do this easily by projecting the image onto the paper. When it's time for the investigation, trace the illustration with markers as you introduce the words and phrases to students. Students will be amazed by your artistic ability.

Science Vocabulary Strategies
- *Word wall*
- *Drawings and diagrams*
- *Cloze activity*
- *Word wizard*
- *Word analysis/word parts*
- *Breaking apart words*
- *Possible sentences*
- *Reading*
- *Glossary*
- *Index*
- *Poems, chants, and songs*

Cloze activity. Structure sentences for students to complete, leaving out the vocabulary words. This can be done as a warm-up with the words from the previous day's lesson. Here's an example from the **Earth History Course**.

>Teacher: *The removal and transportation of loose earth materials is called _____.*
>
>Students: *Erosion.*

Word wizard. Tell students that you are going to lead a word activity. You will be thinking of a science vocabulary word. The goal is to figure out the word. Provide hints that have to do with parts of a definition, root word, prefix, suffix, and other relevant components. Students work in teams of two to four. Provide one hint, and give teams 1 minute to discuss. One team member writes the word on a piece of paper or on the whiteboard, using dark marking pens. Each team holds up its word for only you to see. After the third clue, reveal the word, and move on to the next word. Here's an example.

1. *Part of the word means green.*
2. *They are found in plant cells.*
3. *They look like tiny green spheres or ovals.*

The word is **chloroplasts**.

Word analysis/word parts. Learning clusters of words that share a common origin can help students understand content-area texts and connect new words to familiar ones. Here's an example: *geology, geologist, geological, geography, geometry, geophysical*. This type of contextualized teaching meets the immediate need of understanding an unknown word while building generative knowledge that supports students in figuring out difficult words for future reading.

Breaking apart words. Have teams of two to four students break a word into prefix, root word, and suffix. Give each team different words, and have each team share the parsed elements of the word with the whole class. Here's an example.

>*photosynthesis*
>
>Prefix = *photo*: meaning light
>
>Root = *synthesis*: meaning to put together

Science-Centered Language Development in Middle School

Science-Centered Language Development in Middle School

Possible sentences. Here is a simple strategy for teaching word meanings and generating class discussion.

1. Choose six to eight key concept words from the text of an article in *FOSS Science Resources*.

2. Choose four to six additional words that students are more likely to know something about.

3. Put the list of ten to fourteen words on the board or project it. Provide brief definitions as needed.

4. Ask students to devise sentences that include two or more words from the list.

5. On chart paper, write all sentences that students generate, both coherent and otherwise.

6. Have students read the article from which the words were extracted.

7. Revisit students' sentences, and discuss whether the sentences are sensible based on the passage or how they could be modified to be more coherent.

Reading. After the active investigation, students continue to develop their understanding of the vocabulary words and the concepts those words represent by listening to you read aloud, reading with a partner, or reading independently. Use strategies discussed in the Reading Domain section to encourage students to articulate their thoughts and practice the new vocabulary.

Glossary. Emphasize the vocabulary words students should be using when they answer the focus question in their science notebooks. The glossary in *FOSS Science Resources* or on FOSSweb can be used as a reference.

Index. Have students create an index at the back of their notebooks. There they can record new vocabulary words and the notebook page where they defined and used the new words for the first time in the context of the investigation.

Poems, chants, and songs. As extensions or homework assignments, ask students to create poems, raps, chants, or songs, using vocabulary words from the investigation.

▶ **NOTE**
See the Science Notebooks in Middle School chapter for an example of an index.

ENGLISH-LANGUAGE DEVELOPMENT

Active investigations, together with ample opportunities to develop and use language, provide an optimal learning environment for English learners. This section highlights opportunities for English-language development (ELD) in FOSS investigations and suggests other best practices for facilitating both the learning of new science concepts and the development of academic literacy. For example, the hands-on structure of FOSS investigations is essential for the conceptual development of science content knowledge and the habits of mind that guide and define scientific and engineering practices. Students are engaged in concrete experiences that are meaningful and that provide a shared context for developing understanding—critical components for language acquisition.

When getting ready for an investigation, review the steps and determine the points where English learners may require scaffolds and where the whole class might benefit from additional language-development supports. One way to plan for ELD integration in science is to keep in mind four key areas: prior knowledge, comprehensible input, academic language development, and oral practice. The ELD chart lists examples of universal strategies for each of these components that work particularly well in teaching science.

▶ **NOTE**
English-language development refers to the advancement of students' ability to read, write, and speak English.

English-Language Development (ELD)	
Activating prior knowledge • Inquiry chart • Circle map • Observation poster • Quick write • Kit inventory	**Using comprehensible input** • Content objectives • Multiple exposures • Visual aids • Supported reading • Procedural vocabulary
Developing academic language • Language objectives • Sentence frames • Word wall, word cards, drawings • Concept maps • Cognitive content dictionaries	**Providing oral practice** • Small-group discussions • Science talk • Oral presentations • Poems, chants, and songs • Teacher feedback

Science-Centered Language Development in Middle School

> **NOTE**
> Language forms and structures are the internal grammatical structure of words and how those words go together to make sentences.

Students acquiring English benefit from scaffolds that support the language forms and functions necessary for the academic demands of the science course, that is, accessing science text, participating in productive oral discourse, and engaging in science writing. The table at the end of this section (starting on page 38) provides a resource to help students organize their thinking and structure their speaking and writing in the context of the science and engineering practices. The table identifies key language functions exercised during FOSS investigations and provides examples of sentence frames students can use as scaffolds.

For example, if students are planning an investigation to learn more about insect structures and behaviors, the language objective might be "Students plan and design an investigation that answers a question about the hissing cockroach's behavior." For students who need support, a sentence frame that prompts them to identify the variables in the investigation would provide language forms and structures appropriate for planning their investigation. As a scaffold, sentence frames can also help them write detailed narratives of their procedure. Here's an example from the table.

> **NOTE**
> The complete table appears at the end of the English-Language Development section starting on page 38.

Language functions	Language objectives	Sentence frames
Planning and carrying out investigations		
Design Sequence Strategize Evaluate	Plan controlled experiments with multiple trials. Identify independent variable and dependent variable. Discuss, describe, and evaluate the methods for collecting data.	To find out _____, I will change _____. I will not change _____. I will measure _____. I will observe _____. I will record the data by _____. First, I will _____, and then I will _____. To learn more about _____, I will need _____ to _____.

C32 Full Option Science System

Activating Prior Knowledge

When an investigation engages a new concept, students first recall and discuss familiar situations, objects, or experiences that relate to and establish a foundation for building new knowledge and conceptual understanding. Eliciting prior knowledge also supports learning by motivating interest, acknowledging culture and values, and checking for misconceptions and prerequisite knowledge. This is usually done in the first steps of Guiding the Investigation in the form of a discussion, presentation of new materials, or a written response to a prompt. The tools outlined below can also be used before beginning an investigation to establish a familiar context for launching into new material.

Circle maps. Draw two concentric circles on chart paper. In the smaller circle, write the topic to be explored. In the larger circle, record what students already know about the subject. Ask students to think about how they know or learned what they already know about the topic. Record the responses outside the circles. Students can also do this independently in their science notebooks.

Activating Prior Knowledge
- *Circle maps*
- *Observation posters*
- *Quick writes*
- *Kit inventories*

An example of a circle map

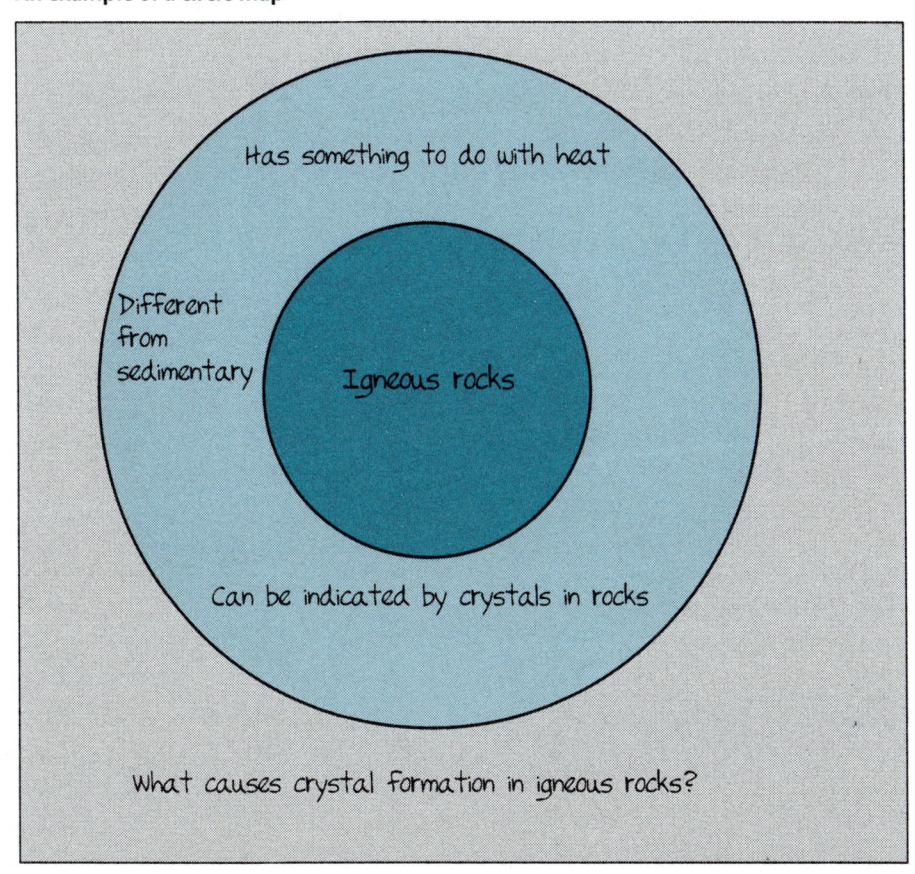

- Has something to do with heat
- Different from sedimentary
- Igneous rocks
- Can be indicated by crystals in rocks
- What causes crystal formation in igneous rocks?

Science-Centered Language Development in Middle School

Science-Centered Language Development in Middle School

Observation posters. Make observation posters by gluing or taping pictures and artifacts relevant to the module or a particular investigation onto pieces of blank chart paper or poster paper. Hang them on the wall in the classroom, and have students rotate in small groups to each poster. At each poster, students discuss their observations with their partners or small groups and then record (write or draw) an observation, a question, a prediction, or an inference about the pictures as a contribution to the commentary on the poster.

As a variation on this strategy, give a set of pictures to each group to pass around. Have them choose one and write what they notice, what they infer, and questions they have in their notebooks.

Quick writes. Ask students what they know about the topic of the investigation. Responses can be recorded independently as a quick write in notebooks and then shared collaboratively. Do not correct misconceptions initially. Periodically revisit the quick-write ideas as a whole class, or have students review their notebook entries to correct, confirm, or complete their original thoughts as they acquire new information (possibly using a line of learning). At the conclusion of the investigation, students should be able to express mastery of the new conceptual material.

Kit inventories. Introduce each item from the FOSS kit used in the investigation, and ask students questions to get them thinking about what each item is and where they may have seen it before. Have them describe the objects and predict how they will be used in the investigation.

Comprehending Input

To initiate their own sense making, students must be able to access the information presented to them. We refer to this ability as comprehending input. Students must understand the essence of new ideas and concepts before beginning to construct new scientific meaning. The strategies for comprehensible input used in FOSS ensure that the instruction is understandable while providing students with the opportunity to grapple with new ideas and the critically important relationships between concepts. Additional tools such as repetition, visual aids, emphasis on procedural vocabulary, and auditory reinforcement can also be used to enhance comprehensible input for English learners.

Content objectives. The focus question for each investigation part frames the activity objectives—what students should know or be able to do at the end of the part. Making the learning objectives clear and explicit prepares English learners to process the delivery of new information, and helps you maintain the focus of the investigation. Write the focus question on the board, have students read it aloud and transcribe it into their science notebooks, and have students answer the focus question at the end of the investigation part. You then check their responses for understanding.

Multiple exposures. Repeat an activity in an analogous but slightly different context, ideally one that incorporates elements that are culturally relevant to students. For example, as a homework assignment for landforms, have students interview their parents about landforms common in the area of their ancestry.

Visual aids. On the board or chart paper, write out the steps for conducting the investigation. This provides a visual reference. Include illustrations if necessary. Use graphic representations (illustrations drawn and labeled in front of students) to review the concepts explored in the active investigations. In addition to the concrete objects in the kit, use realia to augment the activity, to help English learners build understanding and make cultural connections. Graphic organizers (webs, Venn diagrams, T-tables, flowcharts, etc.) aid comprehension by helping students see how concepts are related.

Supported reading. In addition to the reading comprehension strategies suggested in the Reading Domain section of this chapter, English learners can also benefit from methods such as front-loading key words, phrases, and complex text structures before reading; using

---

Comprehending Input
- *Content objectives*
- *Multiple exposures*
- *Visual aids*
- *Supported reading*
- *Procedural vocabulary*

Science-Centered Language Development in Middle School

Science-Centered Language Development in Middle School

Procedural Vocabulary
Add
Analyze
Assemble
Attach
Calculate
Change
Classify
Collect
Communicate
Compare
Connect
Construct
Contrast
Demonstrate
Describe
Determine
Draw
Evaluate
Examine
Explain
Explore
Fill
Graph
Identify
Illustrate
Immerse
Investigate
Label
List
Measure
Mix
Observe
Open
Order
Organize
Pour
Predict
Prepare
Record
Represent
Scratch
Separate
Sort
Stir
Subtract
Summarize
Test
Weigh

preview-review (main ideas are previewed in the primary language, read in English, and reviewed in the primary language); and having students use sentence frames specifically tailored to record key information and/or graphic organizers that make the content and the relationship between concepts visually explicit from the text as they read.

Procedural vocabulary. Make sure students understand the meaning of the words used in the directions for an investigation. These may or may not be science-specific words. Use techniques such as modeling, demonstrating, and body language (gestures) to explain procedural meaning in the context of the investigation. The words students will encounter in FOSS include those listed in the sidebar. To build academic literacy, English learners need to learn the multiple meanings of these words and their specific meanings in the context of science.

Developing Academic Language

As students learn the nuances of the English language, it is critical that they build proficiency in academic language in order to participate fully in the cognitive demands of school. *Academic language* refers to the more abstract, complex, and specific aspects of language, such as the words, grammatical structure, and discourse markers that are needed for higher cognitive learning. FOSS investigations introduce and provide opportunities for students to practice using the academic vocabulary needed to access and engage with science ideas.

Language objectives. Consider the language needs of English learners and incorporate specific language-development objectives that will support learning the science content of the investigation, such as a specific way to expand use of vocabulary by looking at root words, prefixes, and suffixes; a linguistic pattern or structure for oral discussion and writing; or a reading comprehension strategy. Recording in science notebooks is a productive way to optimize science learning and language objectives. For example, in the **Earth History Course**, one language objective might be "Students will apply techniques for rock observations to compare and contrast sedimentary and igneous rocks. They will discuss and record their observations in their notebooks in an organized manner."

Full Option Science System

Vocabulary development. The Science-Vocabulary Development section in this chapter describes the ways science vocabulary is introduced and developed in the context of an active investigation and suggests methods and strategies that can be used to support vocabulary development during science instruction. In addition to science vocabulary, students need to learn the nonscience vocabulary that facilitates deeper understanding and communication skills. Words such as *release*, *convert*, *beneficial*, *produce*, *receive*, *source*, and *reflect* are used in the investigations and *FOSS Science Resources* and are frequently used in other content areas. Learning these academic-vocabulary words gives students a more precise and complex way of practicing and communicating productive thinking. Consider using the strategies described in the Science-Vocabulary Development section to explicitly teach targeted, high-leverage words that can be used in multiple ways and that can help students make connections to other words and concepts. Sentence frames, word walls, concept maps, and cognitive-content dictionaries are strategies that have been found to be effective with academic-vocabulary development.

Science-Centered Language Development in Middle School

Scaffolds That Support Science and Engineering Practices

Language functions	Language objectives	Sentence frames
Asking questions and defining problems		
Inquire Define a problem	Ask questions to solicit information about phenomena, models, or unexpected results; determine the constraints and criteria of a problem.	I wonder why ____ . What happens when ____? What if ____? What does ____? What can ____? What would happen if ____? How does ____ affect ____? How can I find out if ____? Which ____ is better for ____?
Planning and carrying out investigations		
Design Sequence Strategize Evaluate	Plan controlled experiments with multiple trials. Identify independent variable and dependent variable. Discuss, describe, and evaluate the methods for collecting data.	To find out ____, I will change ____. I will not change ____. I will measure ____. I will observe ____. I will record the data by ____. First, I will ____, and then I will ____. To learn more about ____, I will need ____ to ____.

Language functions	Language objectives	Sentence frames
Planning and carrying out investigations *(continued)*		
Describe	Write narratives using details to record sensory observations and connections to prior knowledge.	I observed/noticed ____. When I touch the ____, I feel ____. It smells ____. It sounds ____. It reminds me of ____, because ____.
Organize Compare Classify	Make charts and tables: use a T-table or chart for recording and displaying data.	The table compares ____ and ____
Sequence Compare	Record changes over time, and describe cause-and-effect relationships.	At first, ____, but now ____. We saw that first ____, then ____, and finally ____. When I ____, it ____. After I ____, it ____.
Draw Label Identify	Draw accurate and detailed representations; identify and label parts of a system using science vocabulary, with attention to form, location, color, size, and scale.	The diagram shows ____. ____ is shown here. ____ is ____ times bigger than ____. ____ is ____ times smaller than ____.
Analyzing and interpreting data		
Enumerate Compare Represent	Use measures of variability to analyze and characterize data; decide when and how to use bar graphs, line plots, and two-coordinate graphs to organize data.	The mean is ____. The median is ____. The mode is ____. The range is ____. The x-axis represents ____ and the y-axis represents ____. The units are expressed in ____.

Science-Centered Language Development in Middle School

Science-Centered Language Development in Middle School

Language functions	Language objectives	Sentence frames
Analyzing and interpreting data *(continued)*		
Compare Classify Sequence	Use graphic organizers and narratives to express similarities and differences, to assign an object or action to the category or type to which it belongs, and to show sequencing and order.	This ____ is similar to ____ because ____. This ____ is different from ____ because ____. All these are ____ because ____. ____, ____, and ____ all have/are ____.
Analyze	Use graphic organizers, narratives, or concept maps to identify part/whole or cause-and-effect relationships. Express data in qualitative terms such as more/fewer, higher/lower, nearer/farther, longer/shorter, and increase/decrease; and quantitatively in actual numbers or percentages.	The ____ consists of ____. The ____ contains ____. As ____, then ____. When I changed ____, then ____ happened. The more/less ____, then ____.
Developing and using models		
Represent Predict Explain	Construct and revise models to predict, represent, and explain.	If ____, then ____, therefore ____. The ____ represents ____. ____ shows how ____. You can explain ____ by ____.

Language functions	Language objectives	Sentence frames
Using mathematics and computational thinking		
Symbolize Measure Enumerate Estimate	Use mathematical concepts to analyze data.	The ratio of ____ is ____ to ____. The average is ____. Looking at ____, I think there are ____. My prediction is ____.
Constructing explanations and designing solutions		
Infer Explain	Construct explanations based on evidence from investigations, knowledge, and models; use reasoning to show why the data are adequate for the explanation or conclusion.	I claim that ____. I know this because ____. Based on ____, I think ____. As a result of ____, I think ____. The data show ____, therefore, ____. I think ____ means ____ because ____. I think ____ happened because ____.
Provide evidence	Use qualitative and quantitative data from the investigation as evidence to support claims. Use quantitative expressions using standard metric units of measurement such as cm, mL, °C.	My data show ____. My evidence is ____. The relationship between the variables is ____. The model of ____ shows that ____.

Science-Centered Language Development in Middle School

Science-Centered Language Development in Middle School

Language functions	Language objectives	Sentence frames
Engaging in argument from evidence		
Discuss Persuade Synthesize Negotiate Suggest	Use oral and written arguments supported by evidence and reasoning to support or refute an argument for a phenomenon or a solution to a problem.	I think ___ because___. I agree/disagree with ___ because_____. What you are saying is _____. What do you think about _____? What if _____? I think you should try ___. Another way to interpret the data is _____.
Critique Evaluate Reflect	Evaluate competing design solutions based on criteria; compare two arguments from evidence to identify which is better.	____ makes more sense because ____. ____ is a better design _____ because it ____. Comparing ___ to ___ shows that _____. One discrepancy is ____. ____ is inconsistent with _____. Another way to determine _____ is to _____. I used to think ____, but now I think ____. I have changed my thinking about ____. I am confused about ____ because ____. I wonder ____.
Obtaining, evaluating, and communicating information		
(This practice includes all functions described in the other practices above.)		

Full Option Science System

REFERENCES

Applebee, A. 1984. "Writing and Reasoning." *Review of Educational Research* 54 (Winter): 577–596.

Bereiter, C. 2002. *Education and Mind in the Knowledge Age.* Hillsdale, NJ: Erlbaum.

Black Hills Special Services Cooperative. 2006. "3-2-1 Strategy." In *On Target: More Strategies to Guide Learning.* Rapid City, SD: South Dakota Department of Education.

Gibbons, P. 2002. *Scaffolding Language, Scaffolding Learning.* Portsmouth, NH: Heinemann.

Graham, S., and M. Herbert. 2010. *Writing to Read: Evidence for How Writing Can Improve Reading.* New York: Carnegie.

Hamilton, G. 2002. *Content-Area Reading Strategies: Science.* Portland, ME: Walch Publishing.

Harvey, S. 1998. *Nonfiction Matters: Reading, Writing, and Research in Grades 3–8.* Portland, ME: Stenhouse.

Harvey, S., and A. Goudvis. 2007. *Strategies That Work: Teaching Comprehension for Understanding and Engagement.* Portland, ME: Stenhouse.

Keeley, P. 2008. *Science Formative Assessment: 75 Practical Strategies for Linking Assessment, Instruction, and Learning.* Thousand Oaks, CA: Corwin Press.

Keene, E., and S. Zimmermann. 2007. *Mosaic of Thought: The Power of Comprehension Strategies.* 2nd ed. Portsmouth, NH: Heinemann.

Keys, C. 1999. *Revitalizing Instruction in Scientific Genres: Connecting Knowledge Production with Writing to Learn in Science.* Athens: University of Georgia.

Lieberman, G. A., and L. L. Hoody. 1998. *Closing the Achievement Gap: Using the Environment as an Integrating Context for Learning.* San Diego, CA: State Education and Environment Roundtable.

National Governors Association Center for Best Practices, Council of Chief State School Officers. 2010. *Common Core State Standards for English Language Arts & Literacy in History/Social Studies, Science, and Technical Subjects.* Washington, DC: National Governors Association Center for Best Practices, Council of Chief State School Officers.

National Research Council. 2012. *A Framework for K–12 Science Education: Practices, Crosscutting Concepts, and Core Ideas.* Committee on a Conceptual Framework for New K–12 Science Education Standards. Board on Science Education, Division of Behavioral and Social Sciences and Education. Washington, DC: The National Academies Press.

> **NOTE**
> For additional resources and updated references, go to FOSSweb.

Science-Centered Language Development in Middle School

Norris, S. P., and L. M. Phillips. 2003. "How Literacy in Its Fundamental Sense Is Central to Scientific Literacy." *Science Education* 87 (2).

Ostlund, K. 1998. "What the Research Says about Science Process Skills: How Can Teaching Science Process Skills Improve Student Performance in Reading, Language Arts, and Mathematics?" *Electronic Journal of Science Education* 2 (4).

Wellington, J., and J. Osborne. 2001. *Language and Literacy in Science Education*. Buckingham, UK: Open University Press.

Winokur, J., and K. Worth. 2006. "Talk in the Science Classroom: Looking at What Students and Teachers Need to Know and Be Able to Do." In *Linking Science and Literacy in the K–8 Classroom*, ed. R. Douglas, K. Worth, and W. Binder. Arlington, VA: NSTA Press.

FOSSweb and Technology

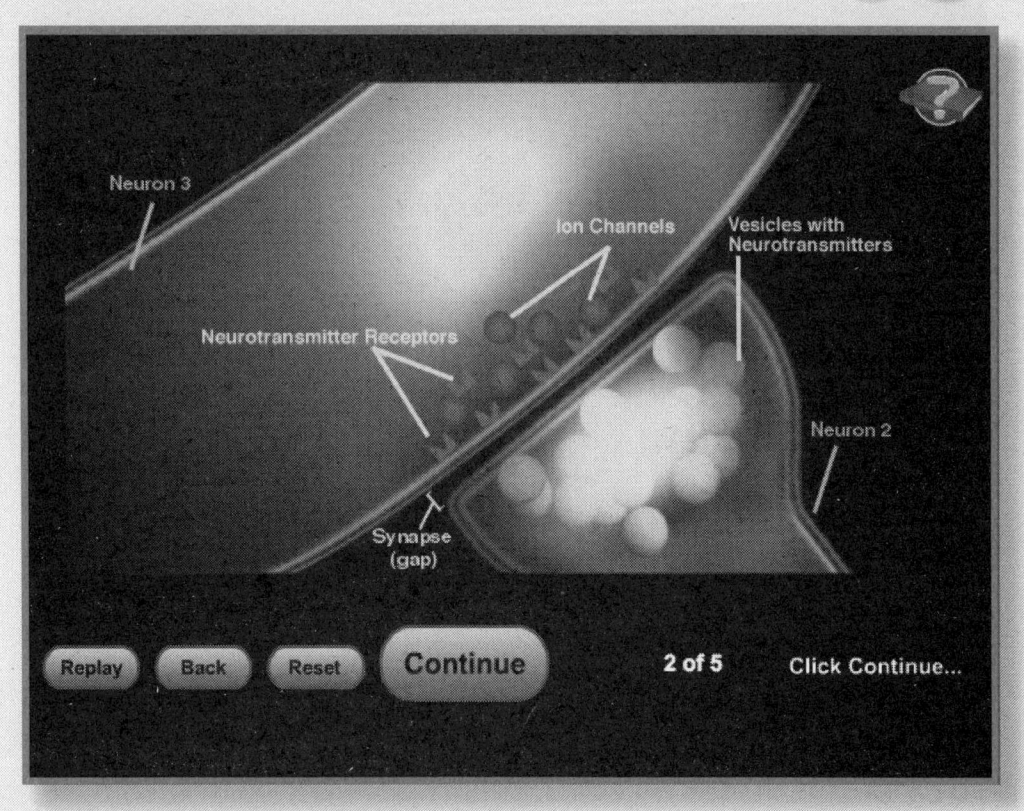

FOSSweb and Technology

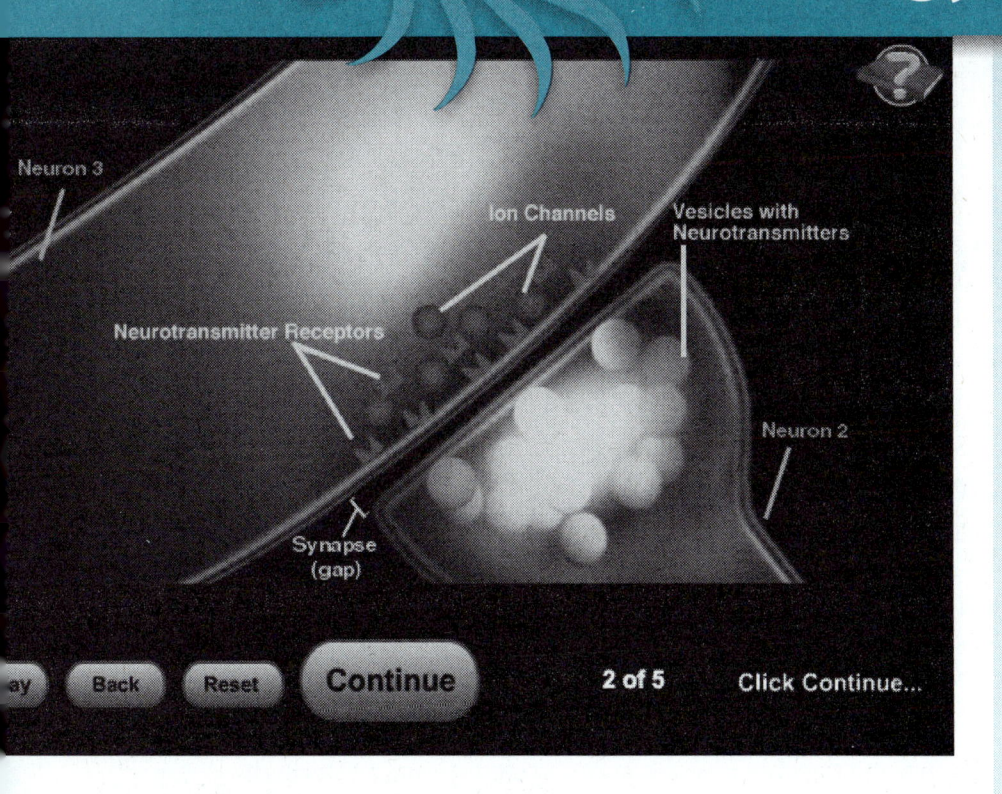

Contents

Introduction D1

Requirements for Accessing
FOSSweb D2

Troubleshooting and
Technical Support D6

INTRODUCTION

FOSSweb technology is an integral part of the **Human Systems Interactions Course**. It provides students with the opportunity to access and interact with simulations, images, video, and text—digital resources that can enhance their understanding of life science concepts. Different sections of digital resources are incorporated into each investigation during the course. Each use is marked with the technology icon in the *Investigations Guide*. You will sometimes use the digital resources to make presentations to the class. At other times, individuals or small groups of students will work with the digital resources to review concepts or reinforce their understanding.

The FOSSweb components are not optional. To prepare to use these digital resources, you should have at a minimum one device with Internet access that can be displayed to the class by an LCD projector with an interactive whiteboard or a large screen arranged for class viewing. Access to a computer lab or to enough devices in your classroom for students to work in small groups is also required during one investigation, and recommended during others.

The digital resources are available online at www.FOSSweb.com for teachers and students. We recommend you access FOSSweb well in advance of starting the course to set up your teacher-user account and become familiar with the resources.

FOSSweb and Technology

REQUIREMENTS FOR ACCESSING FOSSWEB

You'll need to have a few things in place on your device before accessing FOSSweb. Once you're online, you'll create a FOSSweb account. All information in this section is updated as needed on FOSSweb.

Creating a FOSSweb Teacher Account

By creating a FOSSweb teacher account, you can personalize FOSSweb for easy access to the courses you are teaching. When you log in, you will be able to add courses to your "My FOSS Modules" area and access Resources by Investigation for the **Human Systems Interactions Course**. This makes it simple to select the investigation and part you are teaching and view all the digital resources connected to that part.

Students and families can also access course resources through FOSSweb. You can set up a class account and class pages where students will be able to access notes from you about assignments and digital resources.

Setting up an account. Set up an account on FOSSweb so you can access the site when you begin teaching a course. Go to FOSSweb to register for an account—complete registration instructions are available online.

Entering your access code. Once your account is set up, go to FOSSweb and log in. The first time you log in, you will need to enter your access code. Your access code should be printed on the inside cover of your *Investigations Guide*. If you cannot find your FOSSweb access code, contact your school administrator, your district science coordinator, or the purchasing agent for your school or district.

Familiarize yourself with the layout of the site and the additional resources available by using your account login. From your course page, you will be able to access teacher masters, assessment masters, notebook sheets, and other digital resources. Explore the Resources by Investigation section, as this will help you plan. It lists the digital resources, notebook sheets, teacher masters, and readings for each investigation part. There are also a variety of beneficial resources on FOSSweb that can be used to assist with teacher preparation and materials management.

Setting up class pages and student accounts. To enable your students to log in to FOSSweb to access the digital resources and see class assignments, set up a class page and generate a user name and password for the class. To do so, log in to FOSSweb and go to your teacher homepage. Under My Class Pages, follow the instructions to create a new class page and to leave notes for students.

If a class page and student accounts are not set up, students can always access digital resources by visiting FOSSweb.com and choosing to visit the site as a guest.

FOSSweb Technical Requirements

To use FOSSweb, your device must meet minimum system requirements and have a compatible browser and recent versions of any required plug-ins. The system requirements are subject to change. You must visit FOSSweb to review the most recent minimum system requirements.

Preparing your browser. FOSSweb requires a supported operating system with current versions of all required plug-ins. You may need administrator privileges on your device in order to install the required programs and/or help from your school's technology coordinator. Check compatibility for each device you will use to access FOSSweb by accessing the Technical Requirements page on FOSSweb. The information on FOSSweb contains the most up-to-date technical requirements.

https://www.FOSSweb.com/tech-specs-and-info

Support for plug-ins and reader. Any required Adobe plug-ins are available on www.Adobe.com as free downloads. If required, QuickTime is available free of charge from www.Apple.com. FOSS does not support these programs. Please go to the program's website for troubleshooting information.

FOSSweb and Technology

Accessing FOSS Human Systems Interactions Digital Resources

When you log in to FOSSweb, the most useful way to access course materials on a daily basis is the Resources by Investigation section of the **Human Systems Interactions Course** page. This section lists the digital resources, student and teacher sheets, readings, and focus questions for each investigation part. Each of these items is linked so you can click and go directly to that item.

Students will access digital resources from the Resource Room, accessible from the class page you've set up. Explore where the activities reside in the Resource Room. At various points in the course, students will access interactive simulations, images, videos, and animations from FOSSweb.

Other Technology Considerations

Firewall or proxy settings. If your school has a firewall or proxy server, contact your IT administrator to add explicit exceptions in your proxy server and firewall for these servers:

- fossweb.com
- fossweb.schoolspecialty.com
- archive.fossweb.com
- science.video.schoolspecialty.com
- a445.w10.akamai.net

Classroom technology setup. FOSS has a number of digital resources and makes every effort to accommodate users with different levels of access to technology. The digital resources can be used in a variety of ways and can be adapted to a number of classroom setups.

Teachers with classroom devices and an LCD projector, an interactive whiteboard, or a large screen will be able to show multimedia to the class. If you have access to a computer lab, or enough devices in your classroom for students to work in small groups, you can set up time for students to use the FOSSweb digital resources during the school day.

Displaying digital content. You might want to digitally display the notebook and teacher masters during class. In the Resources by Investigation section of FOSSweb, you'll have the option of downloading the masters "to project" or "to copy." Choose "to project" if you plan on projecting the masters to the class. These masters are optimized for a projection system and allow you to type into them while they are displayed. The "to copy" versions are sized to minimize paper use when photocopying for the class, and to fit optimally into student notebooks.

If this projection technology is not available to you, consider making transparencies of the notebook and teacher masters for use with an overhead projector when the Getting Ready section indicates a need to project these sheets.

▶ **NOTE**
FOSSweb activities are designed for a minimum screen size of 1024 × 768. It is recommended that you adjust your screen resolution to 1024 × 768 or higher.

FOSSweb and Technology

TROUBLESHOOTING AND TECHNICAL SUPPORT

If you experience trouble with FOSSweb, you can troubleshoot in a variety of ways.

1. Visit FOSSweb and make sure your devices meet the minimum system requirements.

   https://www.FOSSweb.com/tech-specs-and-info

2. Check the FAQs on FOSSweb for additional information that may help resolve the problem.

3. Try emptying the cache from your browser and/or quitting and relaunching it.

4. Restart your device, and make sure all hardware turns on and is connected correctly.

5. If your school has a firewall or proxy server, contact your IT administrator to add explicit exceptions in your proxy server and firewall for the servers listed on the previous page in this chapter.

If you are still experiencing problems after taking these steps, send FOSS Technical Support an e-mail at support@FOSSweb.com. In addition to describing the problem, include the following information about your device: name of device, operating system, browser and version, plug-ins and versions. This will help us troubleshoot the problem.

Science Notebook Masters

System Summary

1. Which organ system are you researching? _____
2. What are the important functions of this system? List at least three.
3. List several organs found in this system.
4. Choose one organ. Use the online resource "Human Systems Structural Levels" to describe tissues that make up this organ.
5. Choose one tissue that makes up that organ. Use the online resource to describe cells that make up that tissue.
6. What are some general symptoms you might experience if the system you are researching is not functioning properly?

Systems Interactions

Which organ system are you researching? _____

Put a check mark by any organ system your system interacts with. Describe the interaction.

System	√	Interaction description
Muscular		
Skeletal		
Respiratory		
Nervous		
Excretory		
Endocrine		
Digestive		
Circulatory		

Systems Interactions

Which organ system are you researching? _____

Put a check mark by any organ system your system interacts with. Describe the interaction.

System	√	Interaction description
Muscular		
Skeletal		
Respiratory		
Nervous		
Excretory		
Endocrine		
Digestive		
Circulatory		

Connect the Systems

Draw a line from the organ system you researched to all the other systems that it interacts with. Using a different color, do the same for the other systems as you visit each poster.

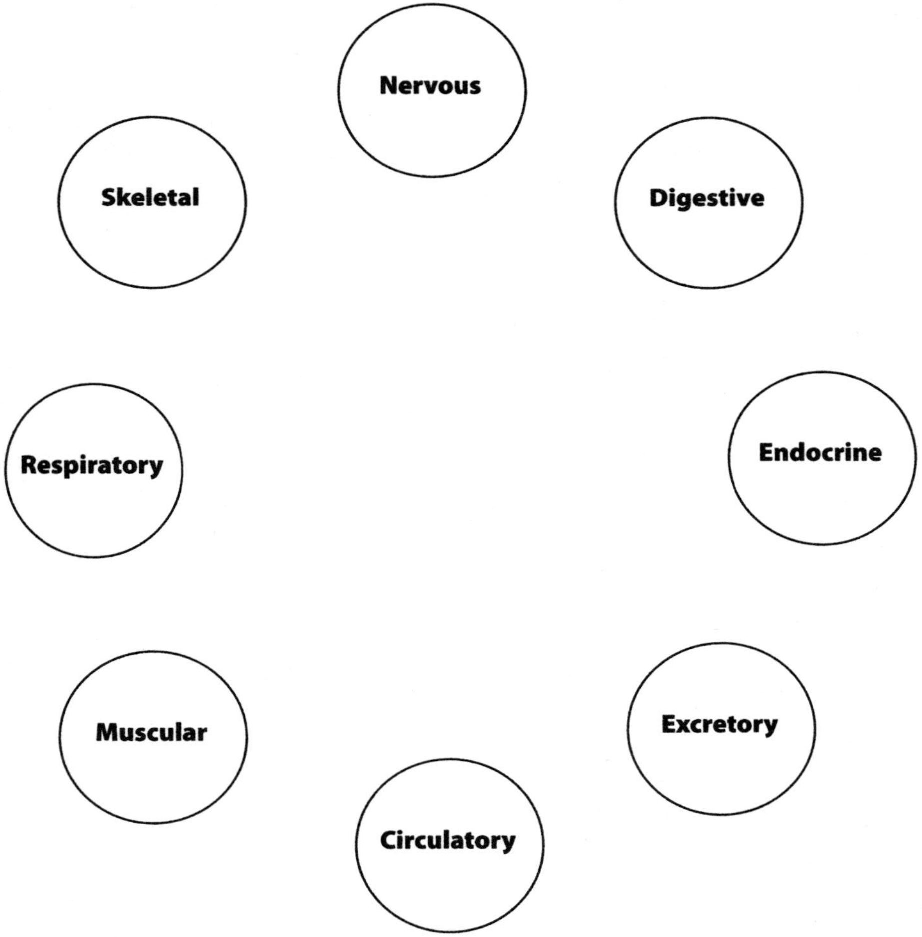

- Skeletal
- Nervous
- Digestive
- Endocrine
- Excretory
- Circulatory
- Muscular
- Respiratory

Connect the Systems

Draw a line from the organ system you researched to all the other systems that it interacts with. Using a different color, do the same for the other systems as you visit each poster.

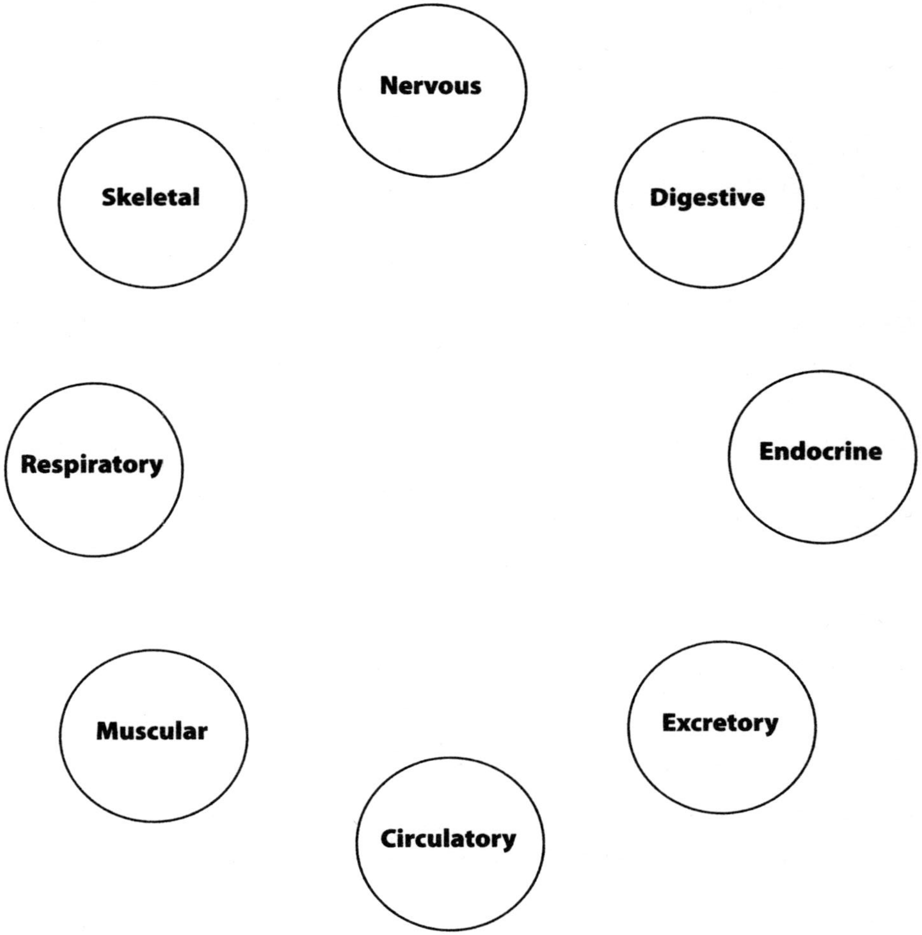

- Skeletal
- Nervous
- Digestive
- Endocrine
- Excretory
- Circulatory
- Muscular
- Respiratory

Organ-Systems Video Questions

Digestive System

1. What is the main function of the digestive system?

2. What happens to food molecules in the villi of the small intestine?

3. The liver separates molecules into toxins (poisons) and nutrients. What happens to the nutrients?

Respiratory System

4. How does oxygen from the air you inhale get into the bloodstream?

5. Where does the bloodstream take the oxygen?

Circulatory System

6. What is the main function of the circulatory system?

Excretory System

7. What is the function of the kidneys in the excretory system?

In your notebook, list at least two systems questions you have.

Response Sheet—Investigation 2

A student was talking with her younger brother about what she was learning. She asked him if he knew how food and oxygen got to the cells in the body. He responded,

Sure, I know. This is how it works: We eat food, and it goes through food tubes to the cells. We breathe in air, and it goes through air tubes to the cells in the body.

Build on the younger brother's ideas to provide a more scientific description of what happens when food and oxygen enter the body and travel to cells.

To and From the Cells

TO CELLS	
FROM CELLS	
AT CELLS (aerobic cellular respiration) $C_6H_{12}O_6 + O_2 \rightarrow Energy + CO_2 + H_2O$	

- Digestive system
- Excretory system
- Circulatory system (blood)
- Respiratory system
- Respiratory system
- Circulatory system (blood)

- Food (glucose)
- Waste water (H_2O)
- Oxygen (O_2)
- Carbon dioxide (CO_2)
- Energy

Touch

Part 1: Braille dots

Follow the instructions to record the number of braille dots you can feel with your fingertip and with your knuckle.

Fingertip

Number of dots											Correct
Y or N											

Knuckle

Number of dots											Correct
Y or N											

What does this tell you about the touch receptors in your fingertips and knuckles?

Part 2: Skin

Label the different touch receptors in the diagram.

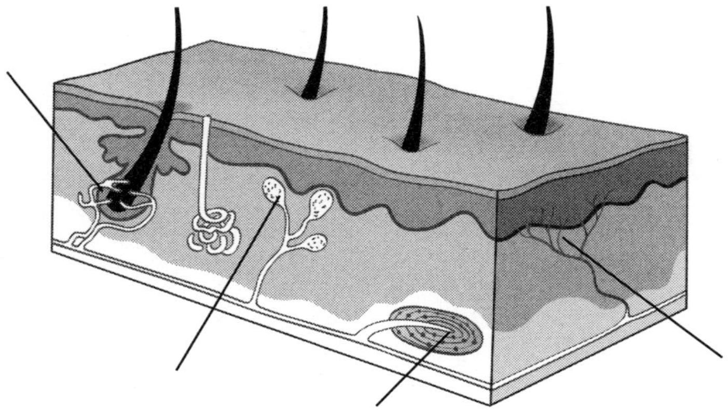

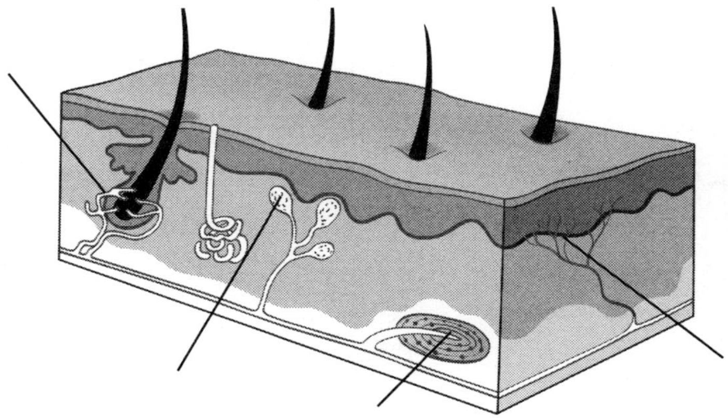

Neural Pathways

Write the function of each structure as you read through "Brain Messages."

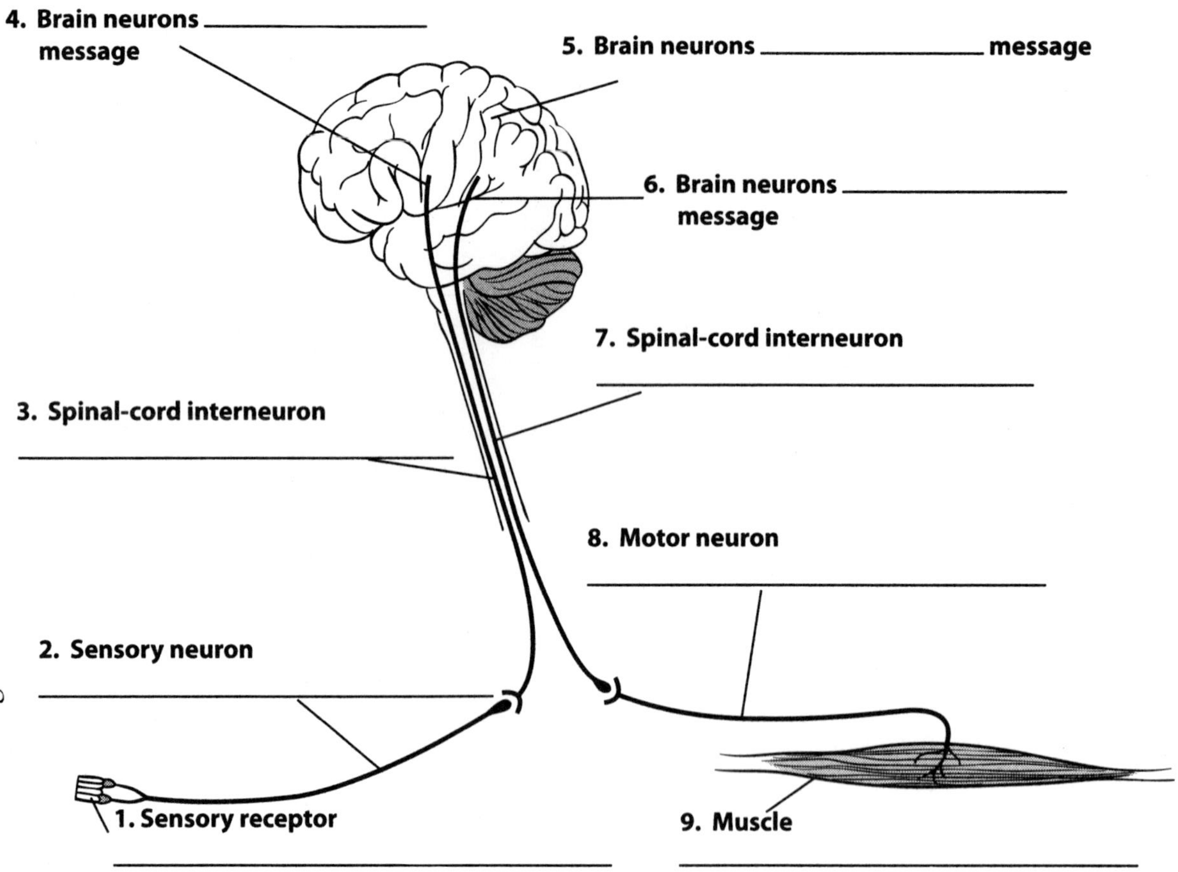

4. Brain neurons _____ message

5. Brain neurons _____ message

6. Brain neurons _____ message

7. Spinal-cord interneuron _____

3. Spinal-cord interneuron _____

8. Motor neuron _____

2. Sensory neuron _____

1. Sensory receptor _____

9. Muscle _____

Neural Pathways

Write the function of each structure as you read through "Brain Messages."

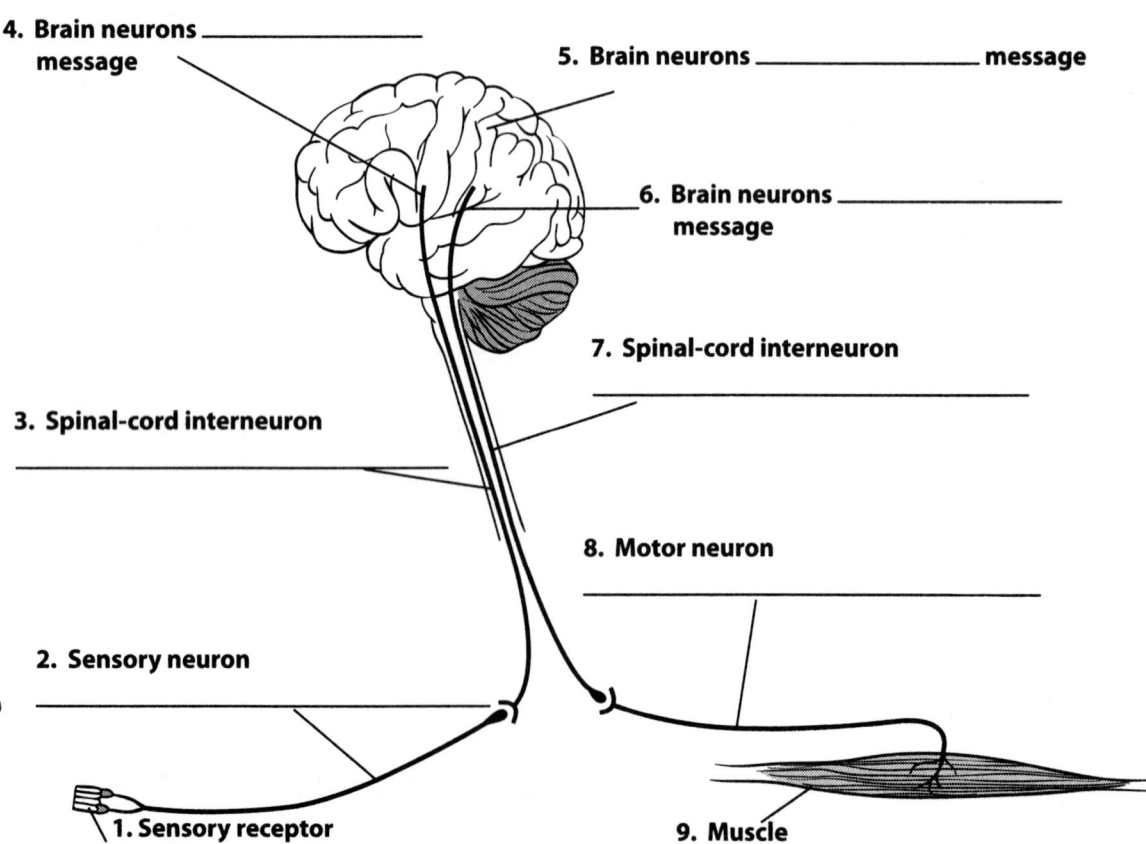

4. Brain neurons _____ message

5. Brain neurons _____ message

6. Brain neurons _____ message

7. Spinal-cord interneuron _____

3. Spinal-cord interneuron _____

8. Motor neuron _____

2. Sensory neuron _____

1. Sensory receptor _____

9. Muscle _____

Response Sheet—Investigation 3

A patient visited the doctor complaining of numbness (lack of feeling) in the fingers on his right hand.

Describe three possible problems in the nervous system that might cause these symptoms. Use a model of the nervous system to explain the phenomena.

Response Sheet—Investigation 3

A patient visited the doctor complaining of numbness (lack of feeling) in the fingers on his right hand.

Describe three possible problems in the nervous system that might cause these symptoms. Use a model of the nervous system to explain the phenomena.

Smell

1. Remove the cap from a numbered vial. Smell the cotton without touching it. Replace the cap.
2. Describe and identify the scent, if you can. Record your observations in the table.

Scent	Description of scent and scent identification: guess	Scent identification: final
1		
2		
3		
4		
5		
6		
7		
8		
Mystery		

Answer these questions in your notebook.

1. Which scents seemed familiar but were difficult to identify? Why do you think they are familiar?
2. Were you able to identify the mystery scent? Why or why not?
3. What kind of receptors are involved in smell?

Reaction-Time Results

1. Your partner will hold *your* reaction timer in front of you.
2. Hold your hand so that your index finger and thumb are approximately 7 centimeters apart. Keep your finger and thumb motionless over the starting position on the reaction timer.
3. Your partner will release the reaction timer. As soon as the timer begins to fall, catch it.
4. Write the number 1 on the reaction timer to show where your finger was when you caught it.
5. Repeat four more times. Number each new catch point: 2, 3, 4, and 5. If you did not catch the timer on an attempt, write that number at the top of the timer.
6. Switch roles.
7. Calculate the reaction times listed below.

My average reaction time	
My group's average reaction time	
Our class's average reaction time	

8. Predict what you think will happen to your reaction time when the room has been darkened.
9. Repeat the process in the dark and record your results below.

My average reaction time	
My group's average reaction time	
Our class's average reaction time	

Answer these questions in your notebook.

1. What happened to the results? Why do you think that happened?
2. What kind of receptors are involved in sight?

Mirror Drawing

Mirror Drawing

"How Memory Works" Video Questions

Answer the following questions as you watch the video.

1. HM drew within the borders of a star just like you did. He did ten trials over 3 days and improved, but had no memory of any prior trials. In other words, he could remember how to perform a skill, but not remember facts or events. What can you conclude about memory from these results? (Remember, part of his brain involved in memory was removed.)

2. One of the scientists remarked, "If you remember anything about this conversation tomorrow, it's because you have a slightly different head (brain) than you had today." Explain what he meant by this statement.

3. What is the importance of molecules (such as PKMzeta) in building long-term memory?

Teacher Masters

AEROBIC CELLULAR RESPIRATION

Teacher Master A

The chemical equation for aerobic cellular respiration:

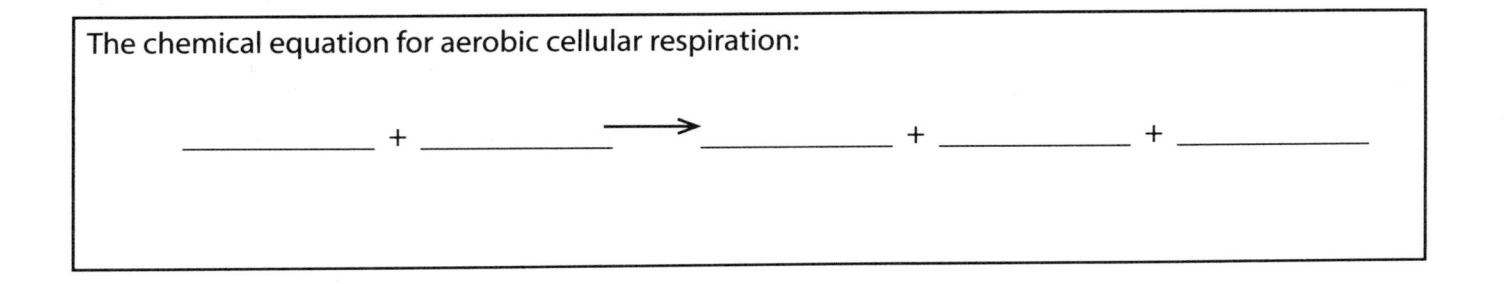

_____ + _____ → _____ + _____ + _____

Blood goes to the excretory system and lungs.

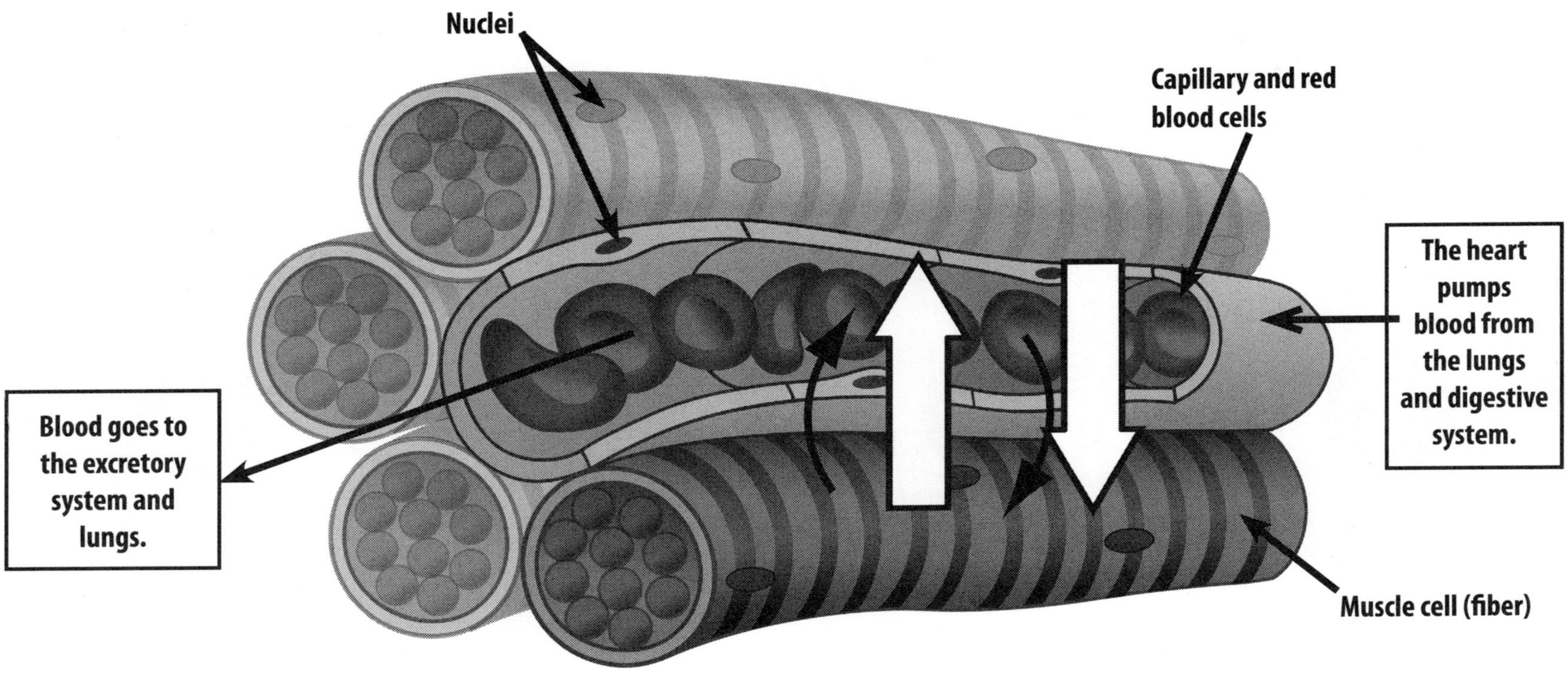

Nuclei

Capillary and red blood cells

The heart pumps blood from the lungs and digestive system.

Muscle cell (fiber)

AEROBIC CELLULAR RESPIRATION TILES

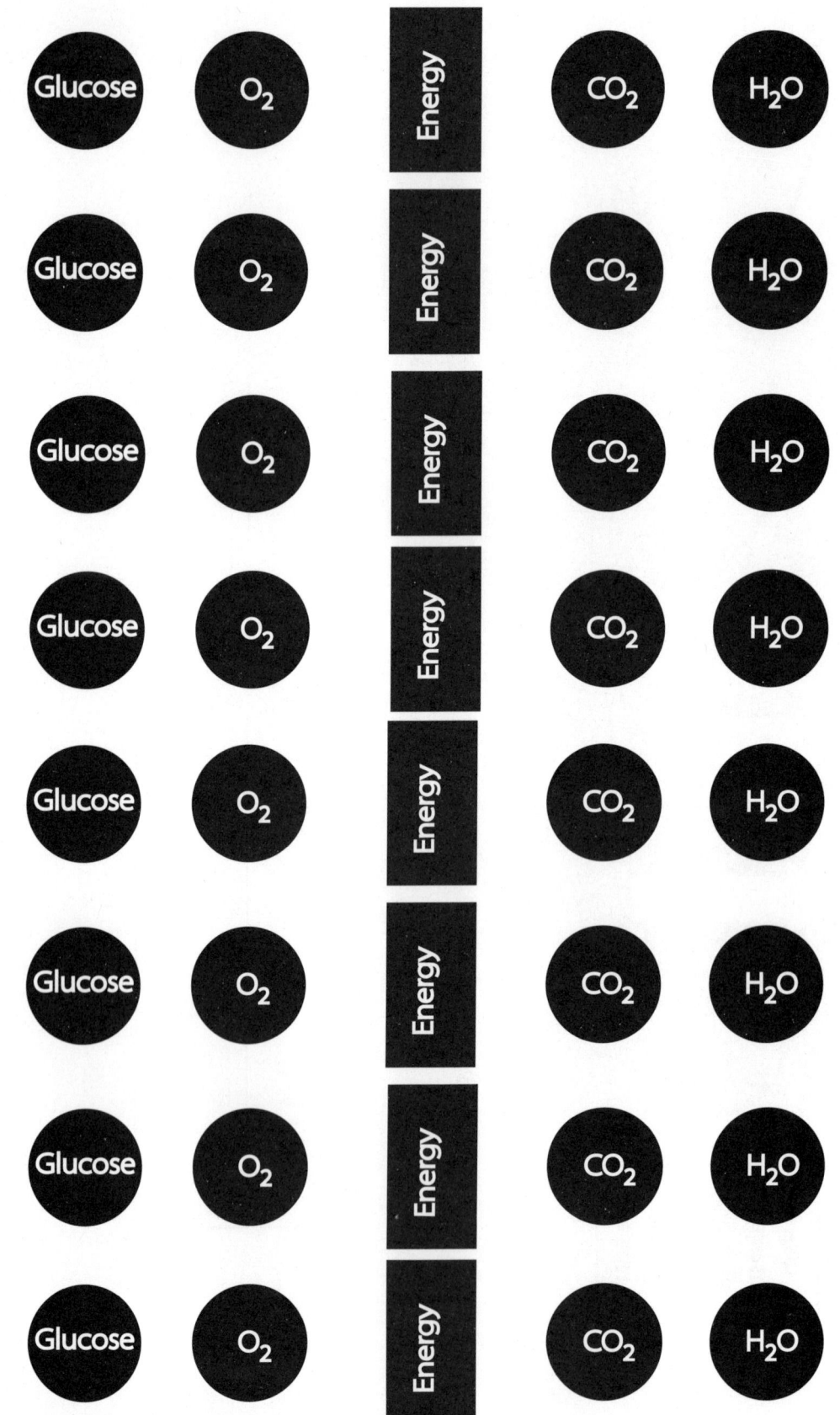

Teacher Master B

(Tiles: 8× Glucose, 8× O₂, 8× Energy, 8× CO₂, 8× H₂O)

FOSS Next Generation
© The Regents of the University of California
Can be duplicated for classroom or workshop use.

Human Systems Interactions Course
Investigation 2: Supporting Cells
Teacher Master B

PRESSURE RECEPTIVE-FIELDS MODELS A

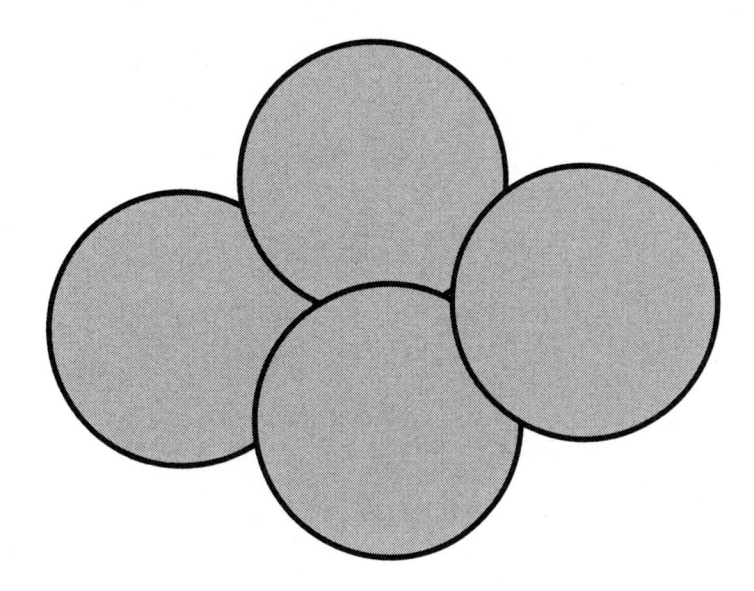

A few large receptive fields

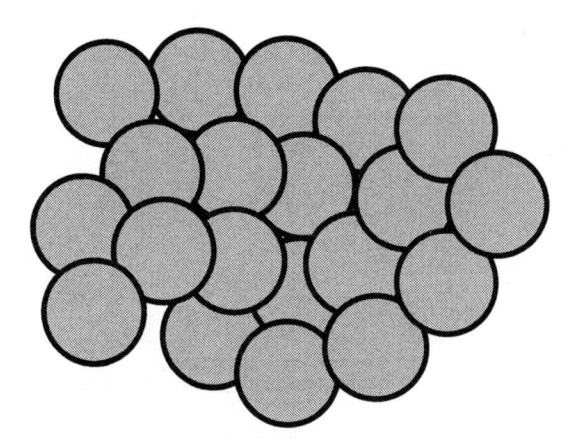

Many small receptive fields

PRESSURE RECEPTIVE-FIELDS MODELS B

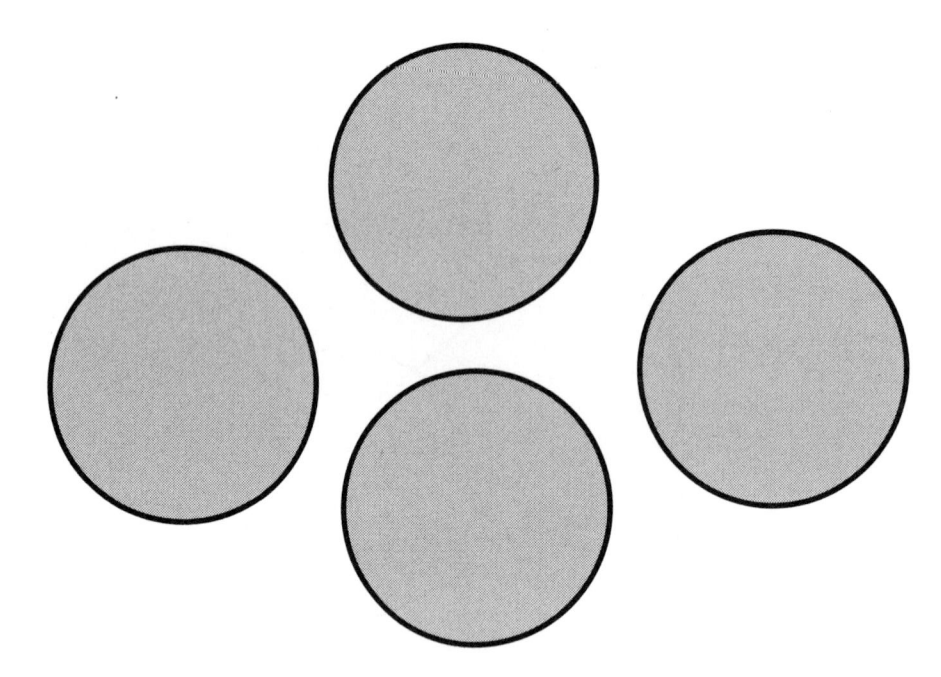

A few large receptive fields

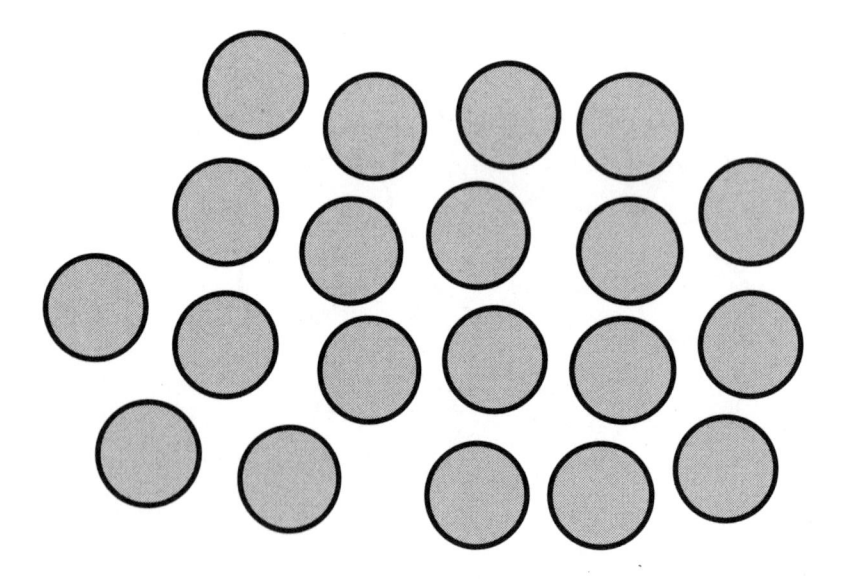

Many small receptive fields

HOMUNCULUS

Teacher Master E

FOSS Next Generation
© The Regents of the University of California
Can be duplicated for classroom or workshop use.

Human Systems Interactions Course
Investigation 3: The Nervous System
Teacher Master E

Teacher Master F

SENDING THE MESSAGE

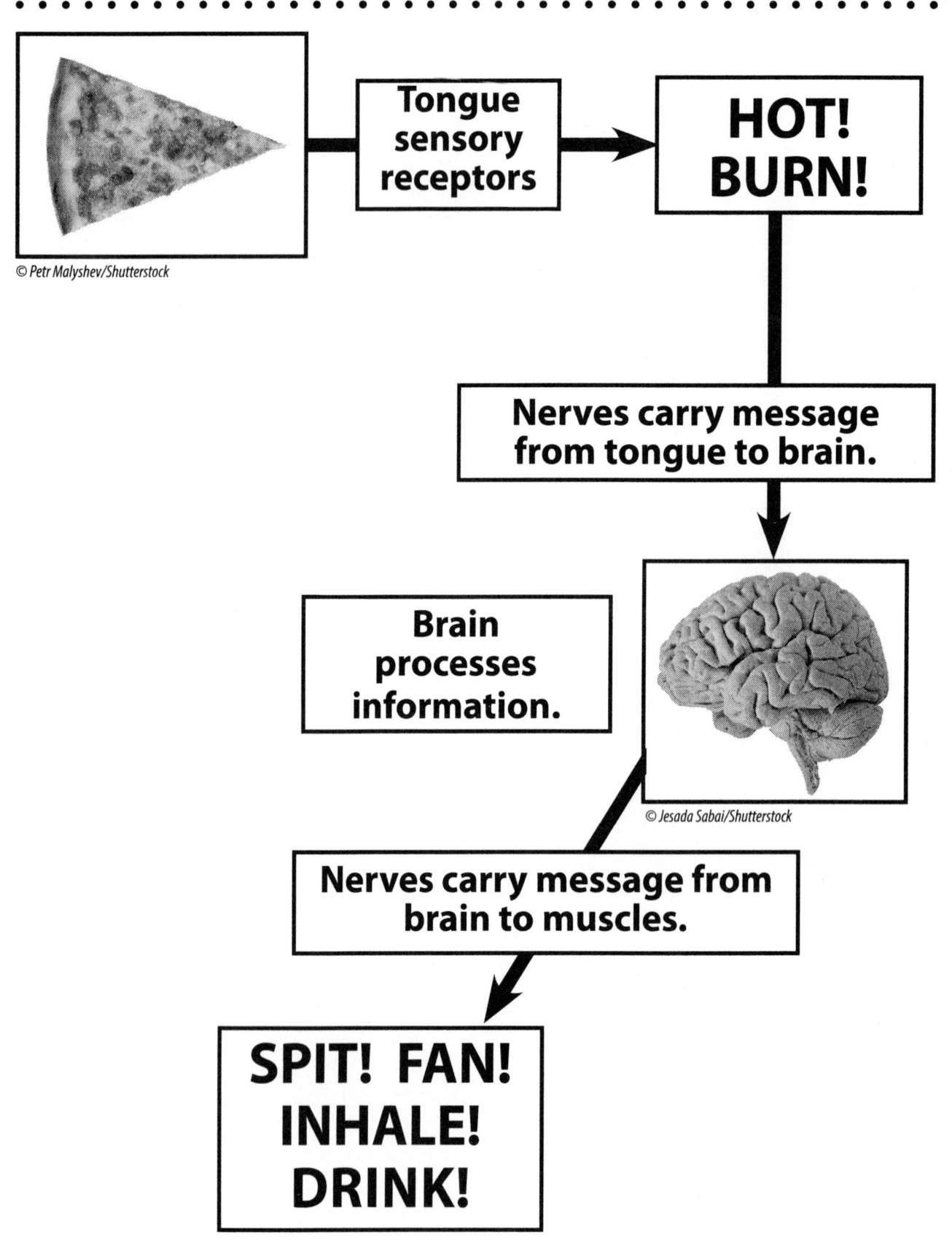

[pizza slice image] © Petr Malyshev/Shutterstock → **Tongue sensory receptors** → **HOT! BURN!**

↓

Nerves carry message from tongue to brain.

↓

Brain processes information. [brain image] © Jesada Sabai/Shutterstock

↓

Nerves carry message from brain to muscles.

↓

SPIT! FAN! INHALE! DRINK!

FOSS Next Generation
© The Regents of the University of California
Can be duplicated for classroom or workshop use.

Human Systems Interactions Course
Investigation 3: The Nervous System
Teacher Master F

NEURAL TRANSMISSION

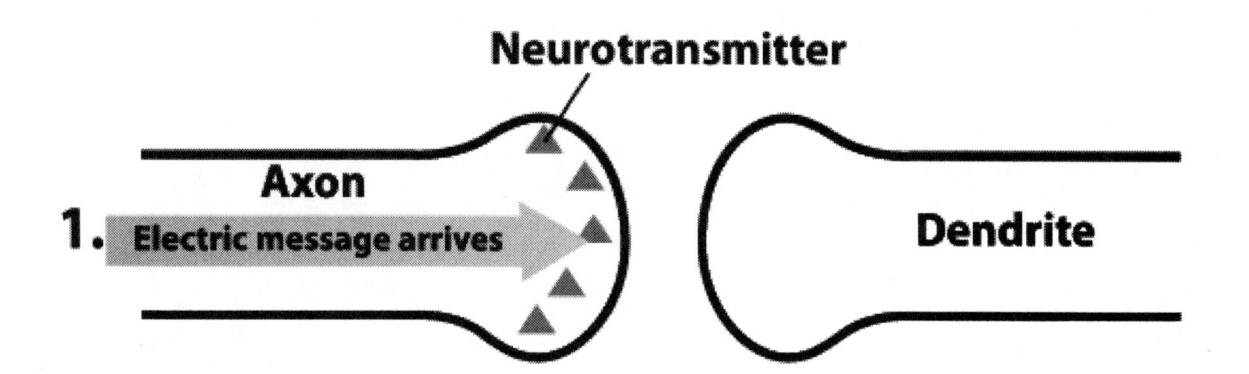

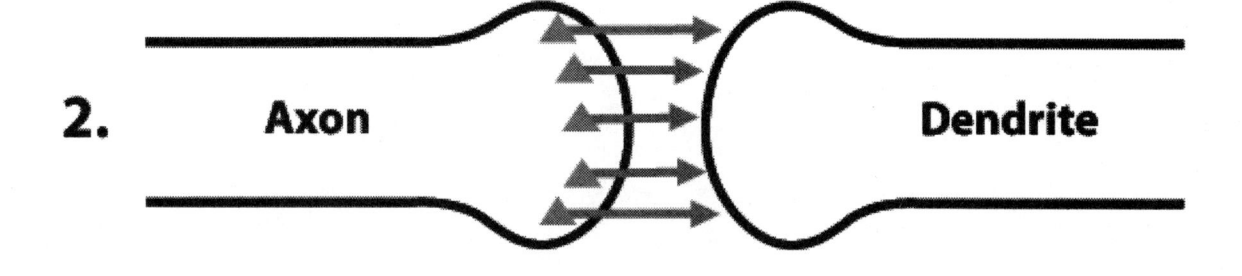

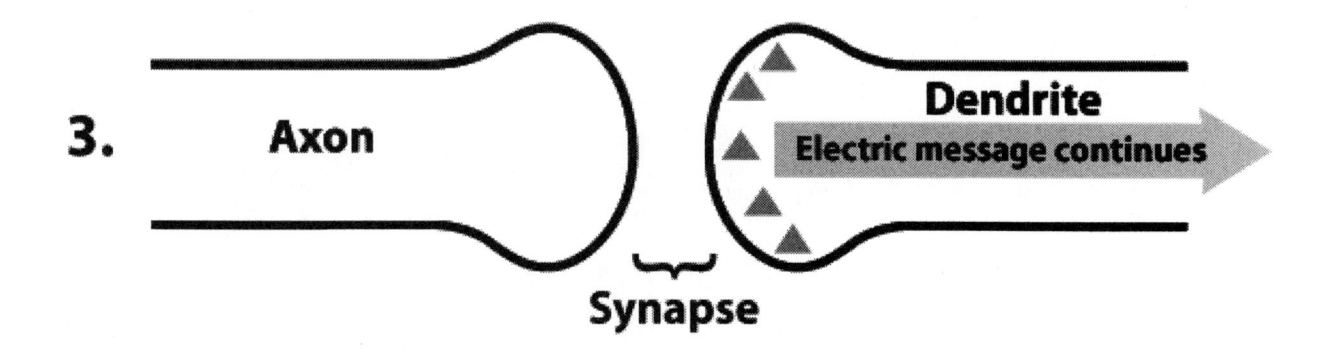

1. Axon — Electric message arrives → Neurotransmitter — Dendrite

2. Axon → → → Dendrite

3. Axon — Dendrite — Electric message continues — Synapse

NEURAL-MESSAGE RELAY CARDS A

PAIN RECEPTOR / SENSORY NEURON

SPINAL-CORD INTERNEURON TO THE BRAIN

BRAIN NEURON RECEIVING MESSAGE

Teacher Master I

NEURAL-MESSAGE RELAY CARDS B

BRAIN NEURON PROCESSING MESSAGE

BRAIN NEURON SENDING MESSAGE

SPINAL-CORD INTERNEURON FROM THE BRAIN

Teacher Master J

NEURAL-MESSAGE RELAY CARDS C

MOTOR NEURON

MUSCLE

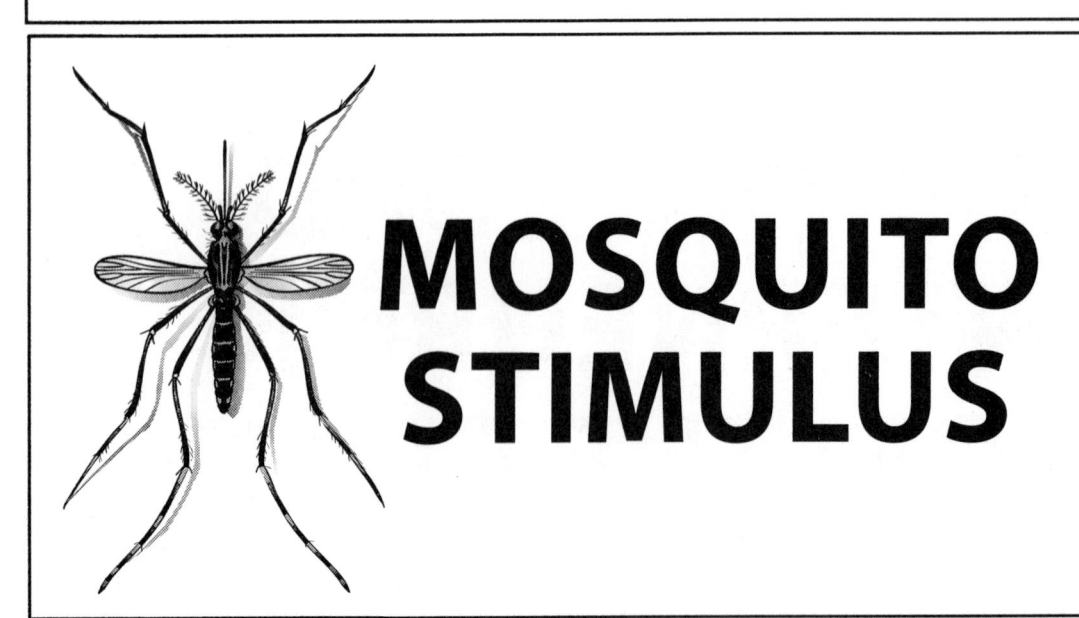

MOSQUITO STIMULUS

FOSS Next Generation
© The Regents of the University of California
Can be duplicated for classroom or workshop use.

Human Systems Interactions Course
Investigation 3: The Nervous System
Teacher Master J

NEURAL-MESSAGE RELAY SETUP

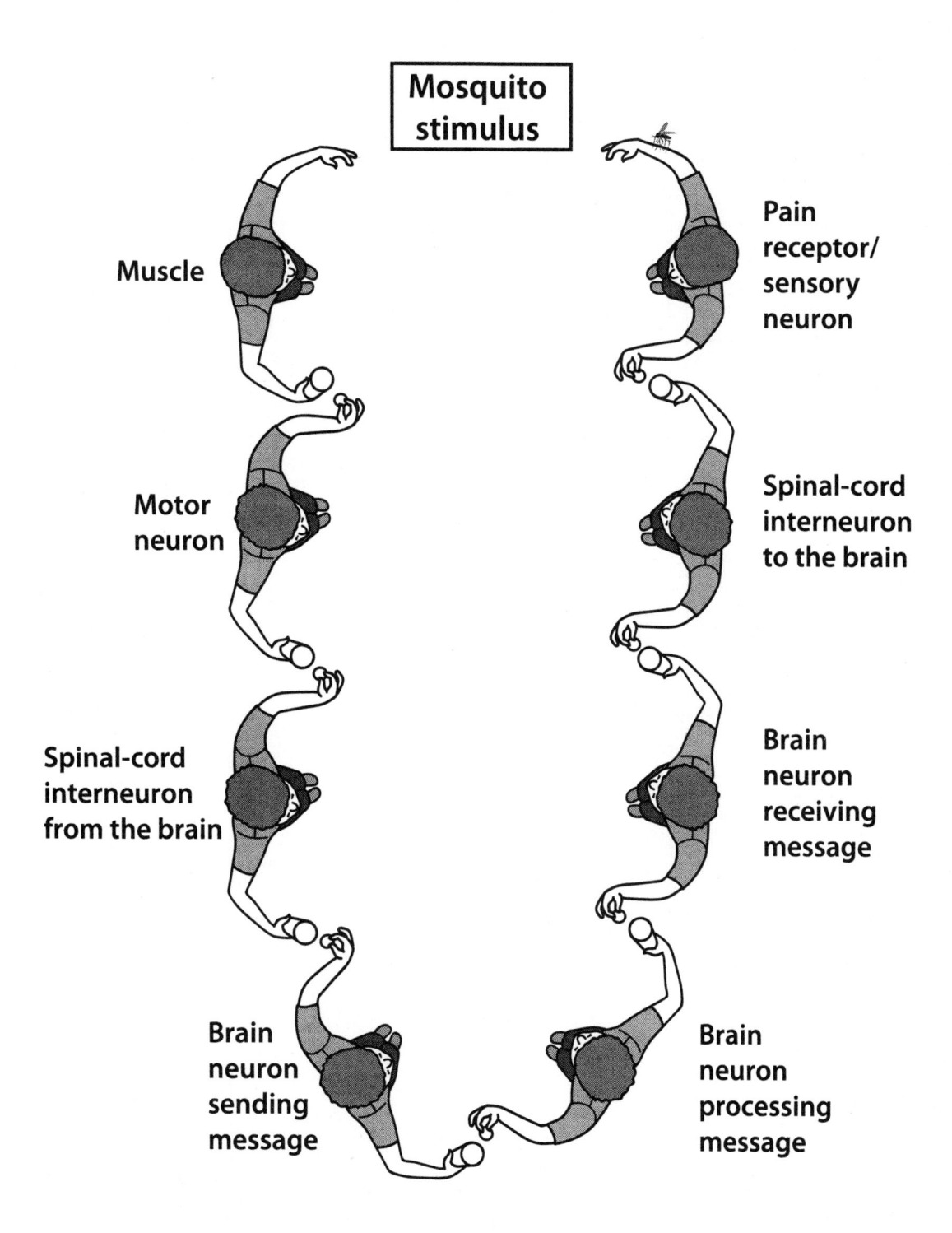

Teacher Master K

Mosquito stimulus

- Muscle
- Motor neuron
- Spinal-cord interneuron from the brain
- Brain neuron sending message
- Pain receptor/sensory neuron
- Spinal-cord interneuron to the brain
- Brain neuron receiving message
- Brain neuron processing message

FOSS Next Generation
© The Regents of the University of California
Can be duplicated for classroom or workshop use.

Human Systems Interactions Course
Investigation 3: The Nervous System
Teacher Master K

REACTION TIMER

Teacher Master L

Reaction time in 100ths of a second	Reaction time in 100ths of a second	Reaction time in 100ths of a second	Reaction time in 100ths of a second
21	21	21	21
20	20	20	20
19	19	19	19
18	18	18	18
17	17	17	17
16	16	16	16
15	15	15	15
14	14	14	14
13	13	13	13
12	12	12	12
STARTING POSITION	**STARTING POSITION**	**STARTING POSITION**	**STARTING POSITION**

FOSS Next Generation
© The Regents of the University of California
Can be duplicated for classroom or workshop use.

Human Systems Interactions Course
Investigation 3: The Nervous System
Teacher Master L

Teacher Master M

REACTION-TIME QUESTIONS

1. What stimulus caused you to respond when the reaction timer dropped?

2. Which photoreceptors were activated in the bright room light?

3. Which photoreceptors were activated when it was dim?

4. Why was it harder to catch the reaction timer in the dim light compared to the bright light?

5. Draw the path the information takes from your eyes to your fingers.

6. How might you change your strategy to improve your reaction time in dim light? Think about where rods are found in the retina.

Teacher Master N

MEMORY SET: HEAR ONLY

1. Tell students to put their pencils down.

2. Read the names out loud. Do not repeat any of the words, but read slowly and loudly.

3. After reading all the words, ask students to list as many objects as they remember in their notebooks.

4. Project the teacher master and have students score their own results. They should circle each correct answer and total them.

salt shaker
coin
zebra
seeds
waffle
toothpick
map
clock
string
bottle

Teacher Master O

MEMORY SET: READ ONLY

1. Tell students to put their pencils down.

2. Project the list of objects.

3. Students read the list silently to themselves.

4. After 30 seconds, cover the teacher master and ask students to write as many objects as they remember.

5. Project the teacher master and have students score their own results. They should circle each correct answer and total them.

measuring cup
camera
eraser
soap
glasses
alligator
pepper
balloon
ring
towel

MEMORY SET: HEAR AND READ

1. Tell students to put their pencils down.

2. Project the list of objects, cover the list, and reveal the objects one by one.

3. Read the names out loud while students read the list silently to themselves.

4. Cover the teacher master and ask students to write as many objects as they remember.

5. Project the teacher master and have students score their own results. They should circle each correct answer and total them.

phone
button
calendar
pen
toothbrush
sheep
watch
hanger
card
nail

Teacher Master Q

MEMORY SET: HEAR, READ, AND WRITE

1. Tell students to get a piece of scratch paper. They can use their pencils for this memory test.

2. Project the list of objects, cover the list, and reveal the objects one by one.

3. Read the names out loud while students read the list silently and write the objects on the scratch paper.

4. Cover the teacher master, have students put the scratch paper away, and ask them to write as many objects as they remember.

5. Project the teacher master and have students score their own results. They should circle each correct answer and total them.

cup
glove
marker
safety pin
ribbon
shell
computer
candy
rubber band
chicken

Teacher Master R

MEMORY SET: HEAR, READ, WRITE, AND SEE OBJECT

1. Tell students to get a piece of scratch paper. They can use their pencils for this memory test.
2. Project the list of objects, cover the list, and reveal the objects one by one.
3. Read the names out loud while students read the list silently and write the objects on the scratch paper.
4. Show students each actual object as you read its name.
5. Cover the teacher master, have students put the scratch paper away, and ask them to write as many objects as they remember.
6. Project the teacher master and have students score their own results. They should circle each correct answer and total them.

key
ruler
paper bag
comb
stapler
dollar bill
book
fork
paper clip
cork

FOSS Next Generation
© The Regents of the University of California
Can be duplicated for classroom or workshop use.

Human Systems Interactions Course
Investigation 3: The Nervous System
Teacher Master R

Assessment Masters

Embedded Assessment Notes

Human Systems Interactions

Investigation ___, Part ___ Date _____

Concept:

Tally: _____ Got it | Doesn't get it _____

Misconceptions/incomplete ideas:

Reflections/next steps:

Investigation ___, Part ___ Date _____

Concept:

Tally: _____ Got it | Doesn't get it _____

Misconceptions/incomplete ideas:

Reflections/next steps:

Investigation ___, Part ___ Date _____

Concept:

Tally: _____ Got it | Doesn't get it _____

Misconceptions/incomplete ideas:

Reflections/next steps:

FOSS Next Generation
© The Regents of the University of California
Can be duplicated for classroom or workshop use.

Human Systems Interactions Course
Embedded Assessment Notes
No. 1—Assessment Master

Performance Assessment Checklist by Group

Human Systems Interactions

Investigation 1, Part 2

Group	Science and Engineering Practices				DCI	Crosscutting Concepts		
	Analyzing and interpreting data	Constructing explanations	Engaging in argument from evidence	Obtaining, evaluating, and communicating information	LS1.A Structure and function	Structure and function	Cause and effect	Systems and system models

NOTE: See the Assessment chapter for a discussion about how to use this checklist.

Performance Assessment Checklist by Student

Human Systems Interactions

Investigation 1, Part 2

Student	Science and Engineering Practices			DCI		Crosscutting Concepts	
	Analyzing and interpreting data	Constructing explanations	Engaging in argument from evidence	Obtaining, evaluating, and communicating information	LS1.A Structure and function	Structure and function	Cause and effect	Systems and system models

NOTE: See the Assessment chapter for a discussion about how to use this checklist.

Performance Assessment Checklist by Group

Human Systems Interactions

Investigation 2, Part 2

Group	Practice	Disciplinary Core Ideas		Crosscutting Concepts
	Developing and using models	LS1.A Structure and function	LS1.C Organization for matter and energy flow in organisms	Systems and system models	Energy and matter

NOTE: See the Assessment chapter for a discussion about how to use this checklist.

FOSS Next Generation
© The Regents of the University of California
Can be duplicated for classroom or workshop use.

Performance Assessment Checklist by Student

Human Systems Interactions

Investigation 2, Part 2

Student	Practice	Disciplinary Core Ideas		Crosscutting Concepts
	Developing and using models	LS1.A Structure and function	LS1.C Organization for matter and energy flow in organisms	Systems and system models	Energy and matter

NOTE: See the Assessment chapter for a discussion about how to use this checklist.

Assessment Record—Entry-Level Survey

Human Systems Interactions Course

Item	Contributes to	Notes for Planning Instruction
1	MS-LS1-1	
2	MS-LS1-3	
3	MS-LS1-7	
4	MS-LS1-7	
5	MS-LS1-8	
6	MS-LS1-8	

Assessment Record—Investigations 1–2 I-Check

Student names	1	2	3	4	5	6	7a	7b	8	9	10

FOSS Next Generation
© The Regents of the University of California
Can be duplicated for classroom or workshop use.

NOTE: A spreadsheet for this chart is available on FOSSweb.com

Assessment Record—Investigation 3 I-Check

Human Systems Interactions Course

Student names	1	2	3	4a-d	4e	5	6	7	8	9

FOSS Next Generation
© The Regents of the University of California
Can be duplicated for classroom or workshop use.

NOTE: A spreadsheet for this chart is available on FOSSweb.com

Assessment Record—Posttest

Human Systems Interactions Course

Student names	1a	1b	2	3	4a	4b	4c	5	6	7	8	9	10

NOTE: A spreadsheet for this chart is available on FOSSweb.com

FOSS Next Generation
© The Regents of the University of California
Can be duplicated for classroom or workshop use.

ENTRY-LEVEL SURVEY
HUMAN SYSTEMS INTERACTIONS

Name _____

Date _____ Class _____

NOTE: If you want to draw a picture to help you answer any of the items on this survey, ask your teacher for a piece of blank paper. Be sure to put your name on that extra piece of paper as well as this one.

1. What is the smallest living system of any organism? Explain why you think so.

   _____
   _____
   _____
   _____

2. In the table below, write the names of as many human body (organ) systems as you can in the column on the left. In the right column, briefly describe what that system does for the body.

Name of system	Function in the human body

3. How do cells get the energy they need to carry out life functions?

   _____
   _____
   _____
   _____

ENTRY-LEVEL SURVEY
HUMAN SYSTEMS INTERACTIONS

Name _____

4. Two students were talking. Student A was examining the fruit salad he was eating for lunch. He said, "Did you know that I am eating energy from the Sun?" Student B was eating a hamburger. She said, "I am eating energy that came from the Sun, too."

    Do you agree with these two students? Explain why you agree or disagree with each student.

    _____
    _____
    _____
    _____
    _____

5. Explain the stimulus-response process if you were to feel an ant crawling on your arm and you use your hand to brush it away.

    _____
    _____
    _____
    _____
    _____

6. How are memories formed?

    _____
    _____
    _____
    _____

INVESTIGATIONS 1–2 I-CHECK
HUMAN SYSTEMS INTERACTIONS

Name _____

Date _____ Class _____

1. Write the letter of the organ system next to a matching function. Use each letter only once.

   _____ gets rid of liquid waste from the body

   _____ transports oxygen to the cells in the body

   _____ gives the body structure and protection

   _____ receives information from the environment

   _____ enables the body to move

   _____ uses hormones to communicate with the body

   _____ turns food into nutrient molecules

   _____ exhales waste gases such as carbon dioxide

   A skeletal system

   B digestive system

   C circulatory system

   D respiratory system

   E excretory system

   F nervous system

   G endocrine system

   H muscular system

2. Describe how organ systems interact to get *oxygen* to the cells.

   _____
   _____
   _____
   _____
   _____

3. Describe how organ systems interact to get *food* to the cells.

   _____
   _____
   _____
   _____
   _____

INVESTIGATIONS 1–2 I-CHECK
HUMAN SYSTEMS INTERACTIONS

Name_____

4. If blood cannot reach heart muscle cells, they can become damaged and die. What might happen to a person if some of his heart muscle cells die?

   Number the answers below to show the order in which structures would be affected if heart muscle cells die. If any phrase is not needed, leave it blank.

   _____ Heart dies

   _____ Circulatory system is unable to function and human dies

   _____ Heart atoms die

   _____ Heart cells die

   _____ Heart tissue dies

5. Explain how the circulatory system supports aerobic cellular respiration.

   _____
   _____
   _____
   _____
   _____
   _____
   _____

6. The one thing that supplies energy to cells is _____.

INVESTIGATIONS 1–2 I-CHECK
HUMAN SYSTEMS INTERACTIONS

Name_____

7. A patient goes to the doctor with the following symptoms: trouble breathing, fatigue, hard time thinking clearly.

   a. Which organ system or systems are most likely affected by the patient's condition?
      (Mark the one best answer.)

      ○ **A** Nervous system only

      ○ **B** Respiratory system only

      ○ **C** Nervous and respiratory systems

      ○ **D** All systems might be affected.

   b. Explain why you chose that answer.

   _____
   _____
   _____
   _____
   _____

8. Aerobic cellular respiration occurs in cell structures called mitochondria. Cells with diseased mitochondria die. Why do those cells die?

   (Mark the one best answer.)

   ○ **A** Aerobic cellular respiration makes food for cells. Without food, the cells die.

   ○ **B** Aerobic cellular respiration makes carbon dioxide and food molecules. The cells becomes full of the extra molecules and die.

   ○ **C** Aerobic cellular respiration provides usable energy for the cells. Without energy, the cells die.

   ○ **D** Aerobic cellular respiration provides cells with oxygen. Without oxygen, the cells die.

INVESTIGATIONS 1–2 I-CHECK
HUMAN SYSTEMS INTERACTIONS

Name_____

9. Mark **X** next to each sentence that helps describe how the *respiratory* system and the *circulatory* system interact in humans.

   _____ The two systems don't interact.

   _____ Blood picks up oxygen in the lungs.

   _____ Blood picks up nutrients from the small intestine.

   _____ Blood carries carbon dioxide to the lungs.

   _____ Blood carries nutrients to cells in the organs of the respiratory system.

   _____ Blood cells are manufactured in the lungs.

10. Aplastic anemia is a disease that decreases the production of red blood cells in the bone marrow of leg and arm bones. One of the symptoms of aplastic anemia is severe headaches. Explain why that might be so.

   _____
   _____
   _____
   _____
   _____
   _____
   _____
   _____
   _____
   _____

INVESTIGATION 3 I-CHECK
HUMAN SYSTEMS INTERACTIONS

Name _____

Date _____ Class _____

1. Write the first letter of the sense next to each phrase it matches. You may use each sense more than once.

   _____ uses chemoreceptors

   _____ allows a person to sense pain

   _____ uses rods and cones

   _____ responds to molecules in the air

   _____ uses photoreceptors

   _____ uses mechanoreceptors

   _____ responds to light

   Key
   T Touch
   S Smell
   V Vision (sight)

2. Spinal injuries can cause loss of sensation and paralysis (loss of ability to move) in the legs. Explain why this can happen.

   _____
   _____
   _____
   _____
   _____
   _____

3. Your lips are more sensitive to touch than your forehead. What is different about the touch receptive fields in the lips compared to the forehead?

   (Mark the one best answer.)

   ○ **A** There are more touch receptive fields in the lips.

   ○ **B** The touch receptive fields are bigger in the lips.

   ○ **C** The touch receptive fields are more sensitive in the lips.

   ○ **D** The touch receptive fields are used more often in the lips.

Name _____

INVESTIGATION 3 I-CHECK
HUMAN SYSTEMS INTERACTIONS

4. Study the image below.

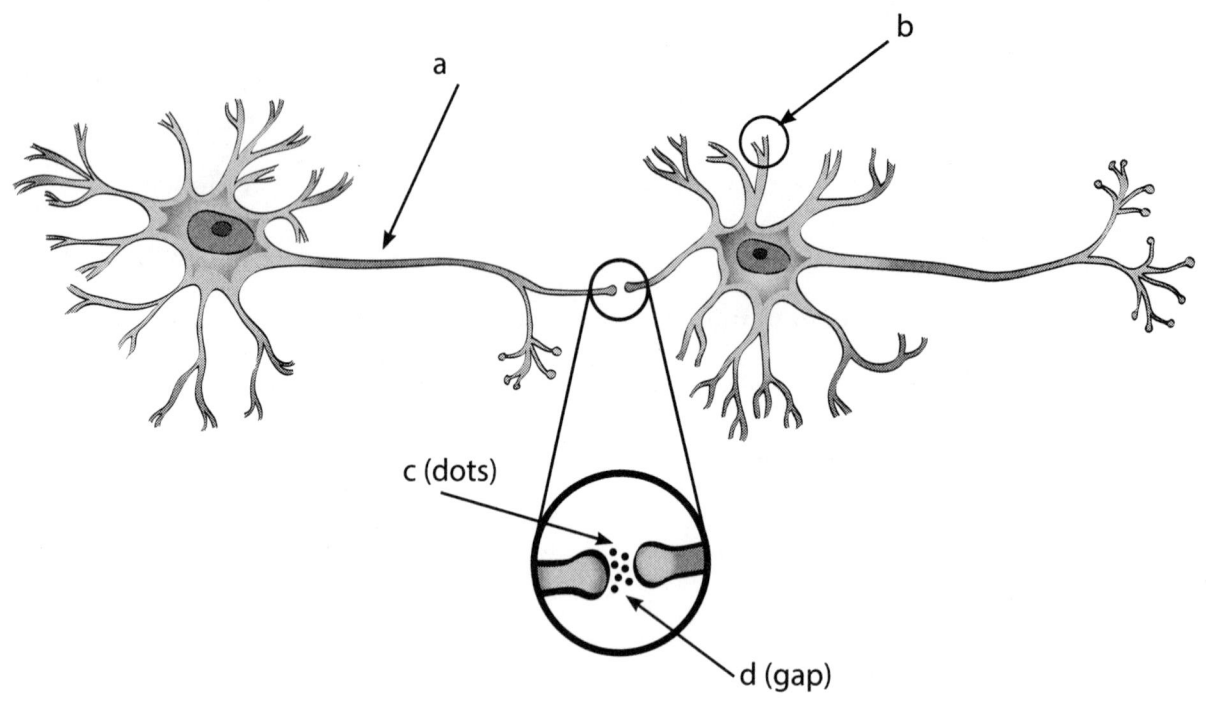

Write the name of each structure on the lines below indicated by a letter in the image above.

a. _____

b. _____

c. _____

d. _____

e. Draw an arrow on the image above that indicates the direction a neural message is transmitted.

5. Neurotransmission has both electric and chemical components. Explain why both are needed.

_____
_____
_____
_____
_____

INVESTIGATION 3 I-CHECK
HUMAN SYSTEMS INTERACTIONS

Name _____

6. A person who is paralyzed can use an electronic device implanted in the brain to control robotic hands. How is that possible?

_____

_____

_____

_____

7. You have an experience that causes a complex network of neurons to be built in your brain. This network is most likely a _____.

    (Mark the one best answer.)

    ○ **A** receptive field

    ○ **B** disease

    ○ **C** memory

    ○ **D** motor pathway

8. Neurotransmitters _____.

    (Mark the one best answer.)

    ○ **A** carry electric impulses along a neuron

    ○ **B** carry chemical impulses along a neuron

    ○ **C** are chemicals that move across synapses

    ○ **D** are chemicals that carry electric impulses across synapses

INVESTIGATION 3 I-CHECK
HUMAN SYSTEMS INTERACTIONS

Name _____

9. Your hand touches something super gooey and icky. In the space below, draw a flowchart or diagram to describe the steps in the neural pathway that leads to pulling your hand out of the goo. Start with the stimulus. Write and draw your answer.

_____
_____
_____
_____
_____
_____
_____

POSTTEST
HUMAN SYSTEMS INTERACTIONS

Name _____

Date _____ Class _____

1. a. Number the following in order from least complex (1) to most complex (4) structure.

    _____ Muscle cell

    _____ Heart

    _____ Heart tissue

    _____ Circulatory system

   b. Number the following in order from least complex (1) to most complex (4) structure.

    _____ Molecule

    _____ Cell

    _____ Atom

    _____ Cell structure (organelle)

2. The smallest structure that could be considered living is the _____.
   (Mark the one best answer.)

    ○ **A** organ

    ○ **B** atom

    ○ **C** molecule

    ○ **D** cell

3. Mark **X** next to each word that describes something that supplies energy to cells.

    _____ Water

    _____ Exercise

    _____ Sleep

    _____ Food

    _____ Caffeine

POSTTEST
HUMAN SYSTEMS INTERACTIONS

Name _____

4. You are taking a test at this moment. Describe how the following are involved.

   a. Photoreceptors

   _____
   _____
   _____

   b. Mechanoreceptors

   _____
   _____
   _____

   c. Memory

   _____
   _____
   _____

5. Scientists have found that chemoreceptors have strong connections to the hippocampus (the part of the brain involved in emotion and memory).

   This means that strong memories or emotions can be stimulated by _____.
   (Mark the one best answer.)

   ○ **A** touch

   ○ **B** smells

   ○ **C** images

   ○ **D** thoughts

POSTTEST
HUMAN SYSTEMS INTERACTIONS

Name _____

6. Memories are _____.
   (Mark the one best answer.)

   ○ **A** complex networks of neurons

   ○ **B** weakened connections between synapses in the brain

   ○ **C** two neurons connected together

   ○ **D** complex electric signals that communicate

7. The muscular and skeletal systems work together to _____.
   (Mark the one best answer.)

   ○ **A** dispose of liquid waste from the body

   ○ **B** pump oxygen into the bloodstream

   ○ **C** transmit information from the environment to the nervous system

   ○ **D** provide structure and protection to the organs of the body

8. • You witness an accident. The injured person is not breathing and has no pulse. The most important action you can take is to call 911, and then _____.
   (Mark the one best answer.)

   ○ **A** give the patient rescue breaths (breathing air into the lungs)

   ○ **B** elevate their legs to send the blood back to the heart

   ○ **C** use chest compressions to keep the heart pumping blood

   ○ **D** try to shake the patient awake and keep him or her talking

   • Why would this be the most important action to take?

   _____

   _____

   _____

POSTTEST
HUMAN SYSTEMS INTERACTIONS

Name _____

9. When scientists refer to *cellular respiration*, they include which of the following processes? *(You may mark as many of the statements below as needed to answer the question.)*

   _____ When air is breathed in and circulated throughout the body

   _____ When plant cells use glucose to release usable energy for the cells

   _____ When usable energy is released in a cell to carry out necessary functions

   _____ When plants transfer water from the ground to all the plant's cells

10. Student A and Student B were having a debate.

    Student A says, "A disease that affects the respiratory system can also affect the nervous system."

    Student B says, "That's not right. If you have allergies, it's harder to breath, but nothing else is affected."

    If you were part of this debate, whose side would you take? Be sure to include evidence that supports the side you want to argue for.

    _____
    _____
    _____
    _____
    _____
    _____
    _____
    _____
    _____
    _____

Notebook Answers

SAMPLE RESPONSES
System Summary

1. Which organ system are you researching? __muscular__
2. What are the important functions of this system? List at least three.
   - moves the bones of the skeletal system
   - moves things like food inside of body
   - causes heart to pump blood
3. List several organs found in this system.

   biceps, deltoids, triceps, epicranius, quadriceps, gastrocnemius
4. Choose one organ. Use the online resource, "Human Systems Structural Levels" to describe tissues that make up this organ.

   quadriceps: striated muscle tissue. It is also made of blood vessels, tendons, and nervous tissue. (I found that in the SRB.)
5. What tissue makes up that organ? Use the online resource to describe cells that make up that tissue.

   Striated muscle tissue is made of skeletal muscle fibers called myocytes.
6. What are some general symptoms you might experience if the system you are researching is not functioning properly?

   weakness, muscle spasms, cramps, twitching, muscle aches, paralysis, difficulty swallowing or breathing, drooping eyelids

---

Systems Interactions

Which organ system are you researching? _____
Put a check mark by any organ system your system interacts with. Describe the interaction.

System	✓	Interaction description
Muscular		
Skeletal		
Respiratory		Answers should describe connections to virtually every system.
Nervous		
Excretory		
Endocrine		
Digestive		
Circulatory		

Organ-Systems Video Questions

Digestive System

1. What is the main function of the digestive system?
   It turns food into nutrient molecules that can be used by the cells.

2. What happens to food molecules in the villi of the small intestine?
   Food molecules pass into the bloodstream.

3. The liver separates molecules into toxins (poisons) and nutrients. What happens to the nutrients?
   Nutrient molecules are carried by the blood to the heart and lungs, and then pumped to all the cells in the body.

Respiratory System

4. How does oxygen from the air you inhale get into the bloodstream?
   Oxygen comes into the lungs into air sacs called alveoli. The alveoli are surrounded by capillaries, and the oxygen passes into the bloodstream.

5. Where does the bloodstream take the oxygen?
   The bloodstream takes oxygen to all the cells in the body.

Circulatory System

6. What is the main function of the circulatory system?
   The circulatory system delivers molecules, such as oxygen and nutrients, to the cells and transports waste molecules to the excretory system.

Excretory System

7. What is the function of the kidneys in the excretory system?
   The kidneys filter out waste.

In your notebook, list at least two systems questions you have.

Connect the Systems

Draw a line from the organ system you researched to all the other systems that it interacts with. Using a different color, do the same for the other systems as you visit each poster.

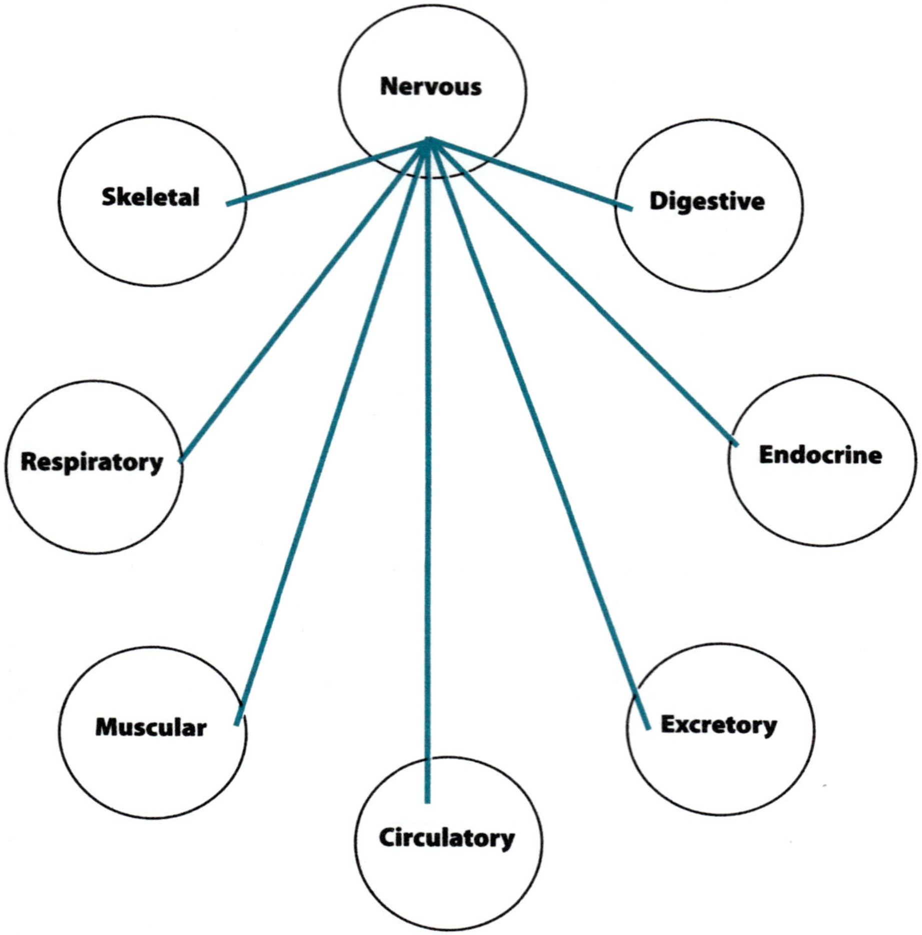

We have shown only the connections for the nervous system. Answers should show lines drawn between virtually every system.

Response Sheet—Investigation 2

A student was talking with her younger brother about what she was learning. She asked him if he knew how food and oxygen got to the cells in the body. He responded,

Sure, I know. This is how it works: We eat food, and it goes through food tubes to the cells. We breathe in air, and it goes through air tubes to the cells in the body.

Build on the younger brother's ideas to provide a more scientific description of what happens when food and oxygen enter the body and travel to cells.

Food enters the body through the mouth, where digestion begins. It transfers to the stomach and then the small intestine, where the food is broken down into nutrient molecules. The nutrient molecules are absorbed into the bloodstream through capillaries and are carried to the cells in the body.

Oxygen enters the body through the mouth and nose and travels to the lungs. It ends up in small air sacs called alveoli, where it is absorbed into the bloodstream through capillaries and travels to the cells in the body.

Blood vessels and the bloodstream carry food and oxygen to the cells. There are no food tubes or air tubes.

This page has been intentionally left blank.

> THIS IS A SAMPLE. YOUR STUDENTS MAY HAVE SOMETHING QUITE DIFFERENT. USE THE WHAT TO LOOK FOR IN THE INVESTIGATIONS GUIDE TO GUIDE YOUR ASSESSMENT.

Food enters the body through the mouth.   Oxygen enters the body through the nose/mouth.

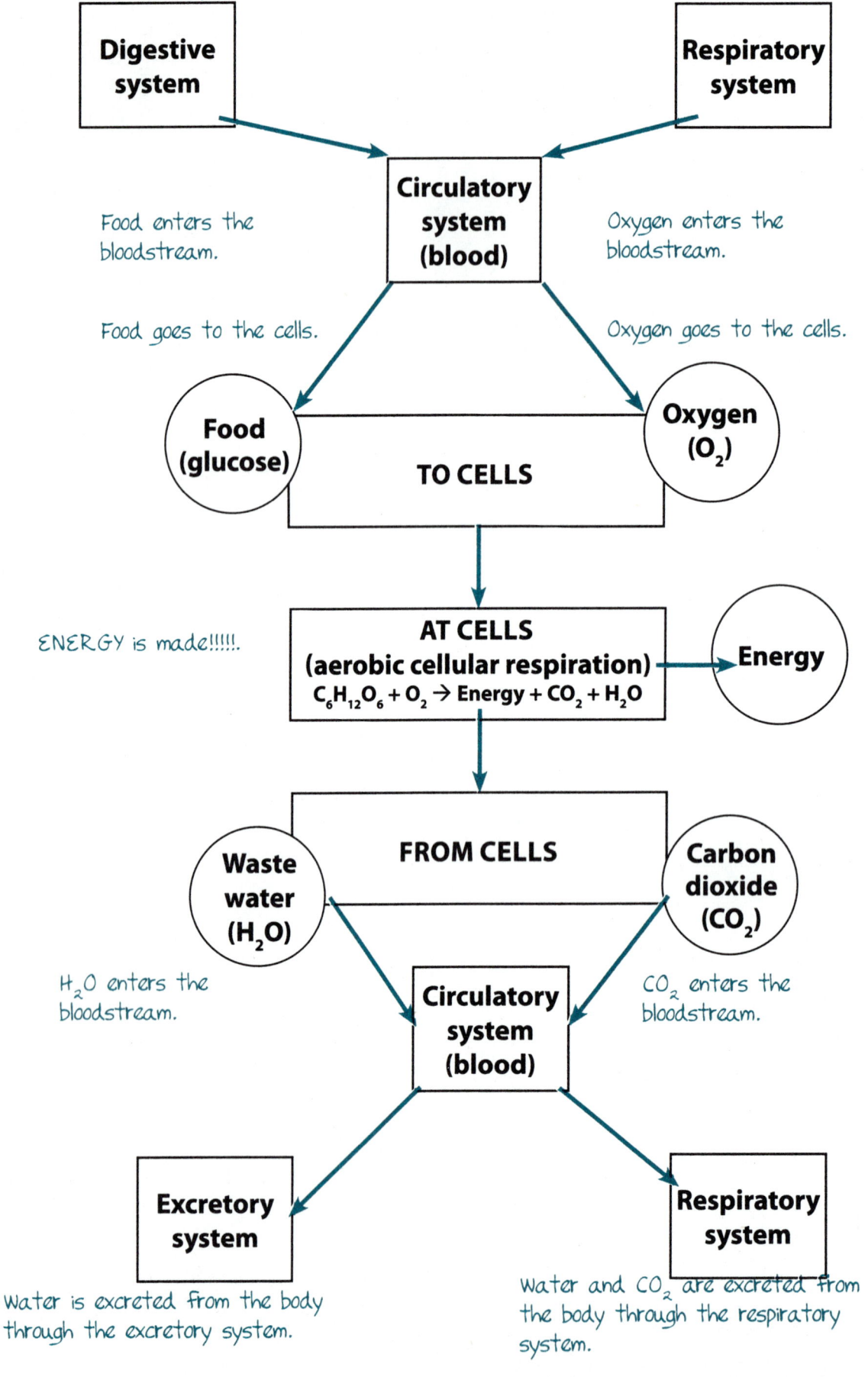

Digestive system → **Circulatory system (blood)** ← **Respiratory system**

Food enters the bloodstream.   Oxygen enters the bloodstream.

Food goes to the cells.   Oxygen goes to the cells.

TO CELLS: Food (glucose), Oxygen (O_2)

ENERGY is made!!!!!

AT CELLS (aerobic cellular respiration)
$C_6H_{12}O_6 + O_2 \rightarrow$ Energy $+ CO_2 + H_2O$

→ Energy

FROM CELLS: Waste water (H_2O), Carbon dioxide (CO_2)

H_2O enters the bloodstream. CO_2 enters the bloodstream.

→ **Circulatory system (blood)** →

Excretory system   **Respiratory system**

Water is excreted from the body through the excretory system.

Water and CO_2 are excreted from the body through the respiratory system.

Touch

Part 1: Braille dots

Follow the instructions to record the number of braille dots you can feel with your fingertip and with your knuckle.

Fingertip															
Number of dots															
Y or N															Correct

Knuckle															
Number of dots															
Y or N															Correct

Answers will vary. Look for completion.

What does this tell you about the touch receptors in your fingertips and knuckles?

The touch receptors have receptive fields that are small. There are many touch receptors and fields; that is why the fingertips are so sensitive. The knuckle has larger and fewer receptive fields.

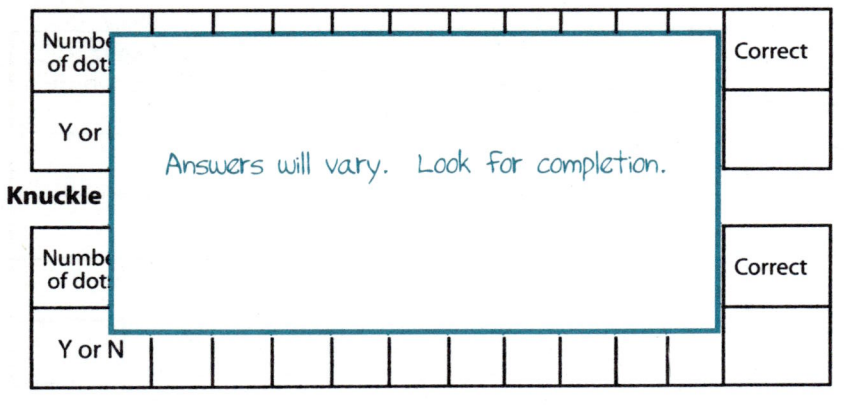

Part 2: Skin

Label the different touch receptors in the diagram.

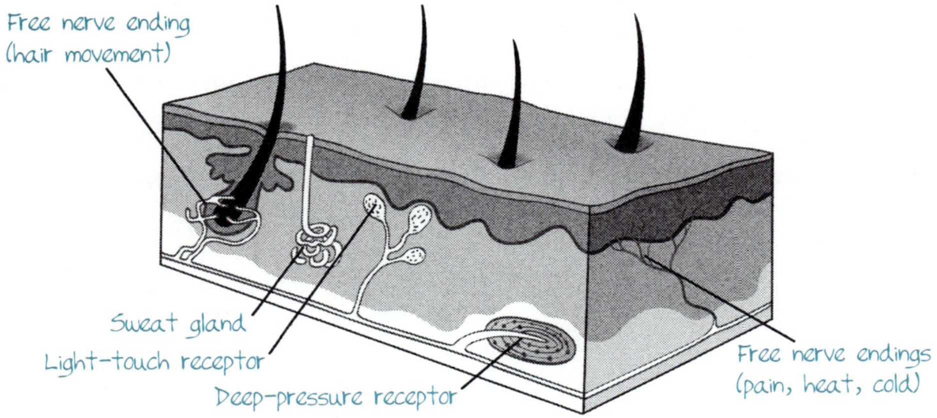

- Free nerve ending (hair movement)
- Sweat gland
- Light-touch receptor
- Deep-pressure receptor
- Free nerve endings (pain, heat, cold)

Neural Pathways

Write the function of each structure as you read through "Brain Messages."

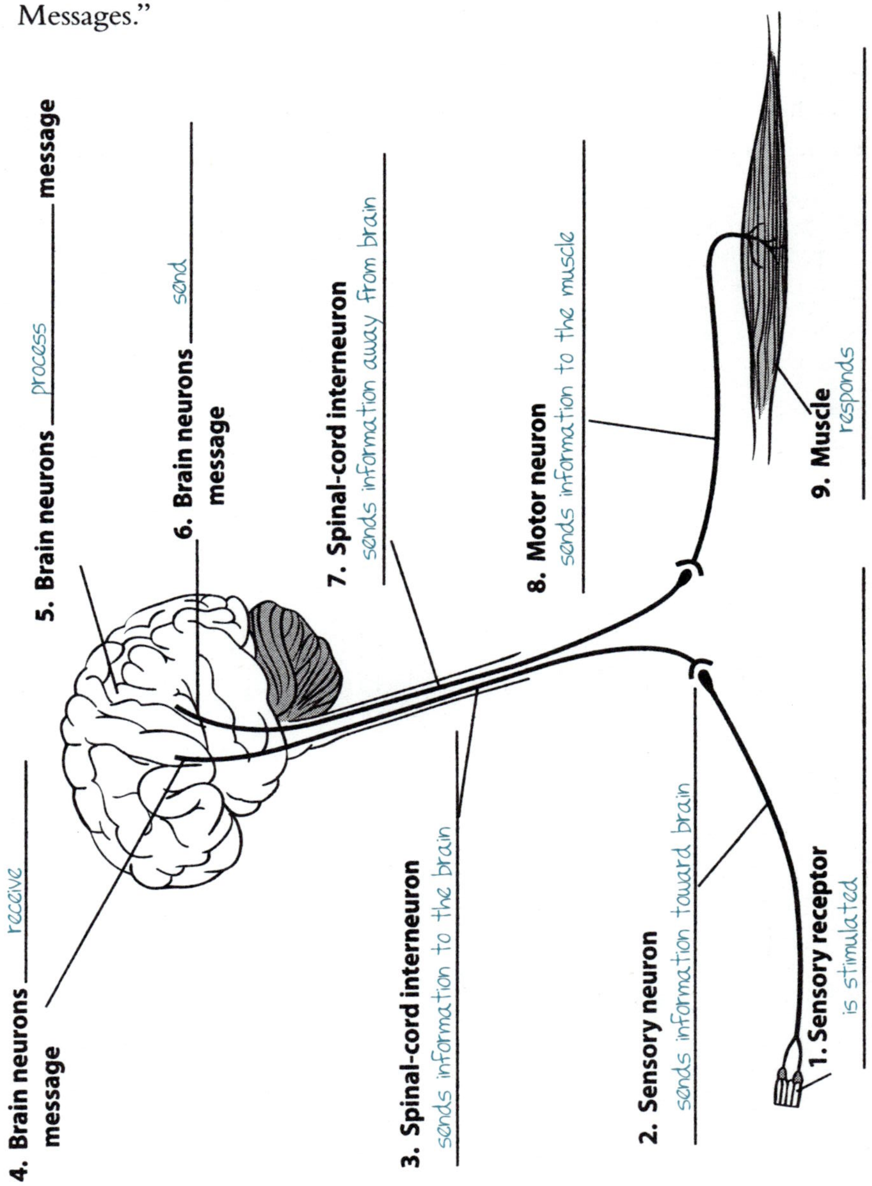

4. Brain neurons __receive__ message

5. Brain neurons __process__ message

6. Brain neurons __send__ message

7. Spinal-cord interneuron
   sends information away from brain

3. Spinal-cord interneuron
   sends information to the brain

8. Motor neuron
   sends information to the muscle

2. Sensory neuron
   sends information toward brain

1. Sensory receptor
   is stimulated

9. Muscle
   responds

Response Sheet—Investigation 3

A patient visited the doctor complaining of numbness (lack of feeling) in the fingers on his right hand.

Describe three possible problems in the nervous system that might cause these symptoms. Use a model of the nervous system to explain the phenomena.

1. Damage to the sensory receptors so they aren't functioning. No signal is getting to the sensory neurons. This could be caused by reduced blood flow to the receptor cells.

2. Damage to the sensory neurons going to the spinal cord and/or to the brain so the message from the receptor is not working. This could be a problem with the chemical neurotransmission process.

3. Problems with the interpretation in the cerebral cortex of the brain and/or the sending of the signal back to the hand.

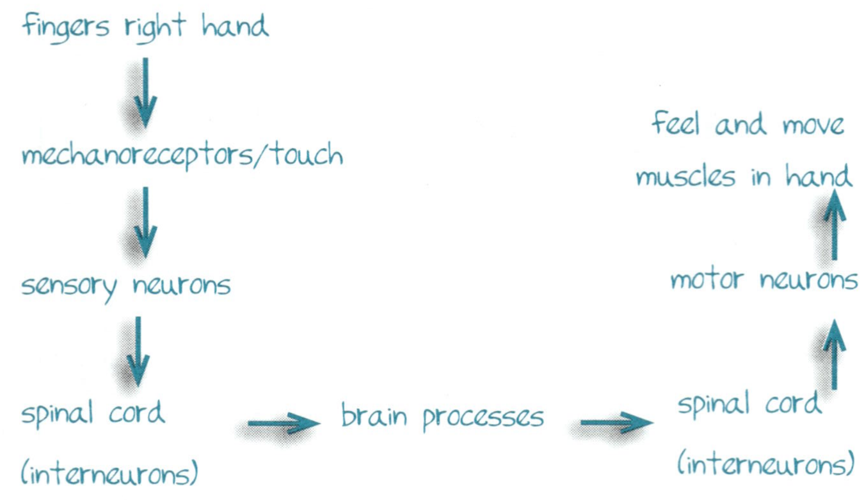

fingers right hand → mechanoreceptors/touch → sensory neurons → spinal cord (interneurons) → brain processes → feel and move muscles in hand ← motor neurons ← spinal cord (interneurons)

Smell

1. Remove the cap from a numbered vial. Smell the cotton without touching it. Replace the cap.
2. Describe and identify the scent, if you can. Record your observations in the table.

Scent	Description of scent and scent identification: guess	Scent identification: final
1		
2		
3	Answers should include thoughtful descriptions and guesses.	
4		
5		
6		
7		
8		
Mystery		

Answer these questions in your notebook.

1. Which scents seemed familiar but were difficult to identify? Why do you think they are familiar? Explanations should relate to experience.
2. Were you able to identify the mystery scent? Why or why not? Explanations should address either confusion with other scents or unfamiliarity with it.
3. What kind of receptors are involved in smell? Chemoreceptors.

Reaction-Timer Results

1. Your partner will hold your reaction timer in front of you.
2. Hold your hand so that your index finger and thumb are approximately 7 centimeters apart. Keep your finger and thumb motionless over the starting position on the reaction timer.
3. Your partner will release the reaction timer. As soon as the timer begins to fall, catch it.
4. Write the number 1 on the reaction timer to show where your finger was when you caught it.
5. Repeat four more times. Number each new catch point: 2, 3, 4, and 5. If you did not catch the timer on an attempt, write that number at the top of the timer.
6. Switch roles.
7. Calculate the reaction times listed below.

My average reaction time	
My group's average reaction time	
Our class's average reaction time	

8. Predict what you think will happen to your reaction time when the room has been darkened.
9. Repeat the process in the dark and record your results below.

My average reaction time	
My group's average reaction time	
Our class's average reaction time	

Answer these questions in your notebook.

1. What happened to the results? Why do you think that happened?
   Explanations should reflect results. Most likely the time increased, which means that students could not see the reaction timer and reacted more slowly to its movement.
2. What kind of receptors are involved in sight?
   Photoreceptors. Rods help see in dim light.

"How Memory Works" Video Questions

Answer the following questions as you watch the video.

1. HM drew within the borders of a star just like you did. He did ten trials over 3 days and improved, but had no memory of any prior trials. In other words, he could remember how to perform a skill, but not remember facts or events. What can you conclude about memory from these results? (Remember, part of his brain involved in memory was removed.)
   Memory must be processed in other parts of the brain besides those that were removed during surgery.

2. One of the scientists remarked, "If you remember anything about this conversation tomorrow, it's because you have a slightly different head (brain) than you had today." Explain what he meant by this statement.
   Memories form when neurons change to increase the network connections at the synapses.

3. What is the importance of molecules (such as PKMzeta) in building long-term memory?
   These molecules appear to glue neurons together, maintaining the networks over time as long-term memories form.